ENERGIZING SUSTAINABLE CITIES

Urban systems now house about half of the world's population, but determine some three-quarters of the global economy and its associated energy use and resulting environmental impacts. The twenty-first century will be increasingly urban. Sustainable development therefore needs first to be defined and analyzed, and then realized in urban settings. Energy is one of the key challenges, but also one of the key opportunities in the required urban sustainability transition.

The book is the result of a major international effort to conduct the first comprehensive assessment of energy-related urban sustainability issues conducted under the auspices of the Global Energy Assessment (GEA). The assessment is also unique in that it embeds energy issues into the broader sustainability agenda of cities: including housing for the poor, functional transport systems, as well as environmental quality, in addition to the challenges imposed by climate change.

Written by an eminent team of internationally renowned scholars, it presents new data, new analysis, as well as new policy insights. It includes the first comprehensive global coverage overview of urban energy use and of the specifics of urban energy demand and supply. Major development and sustainability challenges of cities are assessed in detail and public and private sector opportunities and constraints of a sustainability transition examined. Technological and policy options are put in a much-needed context in terms of their respective roles as drivers of urban energy demand as well as potentials for reductions in energy use and associated emissions of local pollutants as well as greenhouse gases. The analysis presents both a comprehensive literature review as well as novel, spatially explicit models of integrated urban energy policy analysis. The volume concludes with a summary assessment of policy options, priorities as well as paradoxes.

Arnulf Grubler is Acting Program Leader of the Transitions to New Technologies Program at the International Institute for Applied Systems Analysis, Laxenburg, Austria, and Professor in the Field of Energy and Technology at the School of Forestry and Environmental Studies, Yale University.

David Fisk is a co-director of the BP Urban Energy Project, and holds the Laing O'Rourke Chair in Systems Engineering and Innovation at Imperial College London.

ENERGIZING SUSTAINABLE CITIES

Assessing Urban Energy

Edited by
Arnulf Grubler and David Fisk

earthscan
from Routledge

Routledge
Taylor & Francis Group

LONDON AND NEW YORK

First published 2013
by Routledge
2 Park Square, Milton Park, Abingdon, Oxon OX14 4RN

Simultaneously published in the USA and Canada
by Routledge
711 Third Avenue, New York, NY 10017

Routledge is an imprint of the Taylor & Francis Group, an informa business

British Library Cataloguing in Publication Data
A catalogue record for this book is available from the British Library

Library of Congress Cataloging in Publication Data
Energizing sustainable cities : assessing urban energy / edited by Arnulf
Grubler and David Fisk.
p. cm.
Includes bibliographical references and index.
1. Cities and towns--Energy consumption. 2. Energy policy. 3. Energy
conservation--Planning. 4. Sustainable development. I. Gr Arnulf, 1955- II. Fisk,
David (David J.)
HD9502.A2E5355 2012
333.7909173'2--dc23
2012000250

ISBN: 9781849714389 (hbk)
ISBN: 9781849714396 (pbk)
ISBN: 9780203110126 (ebk)

Typeset in 10/12pt Palatino
by Saxon Graphics Ltd, Derby

MIX
Paper from
responsible sources
FSC® C004839
www.fsc.org

Printed and bound in Great Britain by
TJ International Ltd, Padstow, Cornwall

Contents

Plates

Figures

Boxes

Tables

Notes on contributors

Xuemei Bai is Professor of Urban Environment at the Fenner School of Environment and Society, Australian National University. Her educational background is in science and engineering. Her current research focuses on urban sustainability science, including: understanding the structure, function and processes of urban social ecological systems; drivers and impacts of urbanization; urban metabolism; urban environmental management, policy and governance; sustainability transition in Asia; and regional development and environmental policy in China. Most of her work focuses on Asia. She is a Member of the Scientific Committee of the International Human Dimensions Programme for Global Change. She is an adjunct professor at the Institute of Urban Environment, Chinese Academy of Sciences.

Thomas Buettner is Assistant Director and Chief of the Population Studies Branch of the Population Division at the Department of Economic and Social Affairs at the United Nations (UN), New York. A demographer by training, his research focuses on the determinants and methodologies for long-range population projections. At the UN he coordinated and contributed both to the UN's state-of-art world population projections as well as to its urbanization projections; published regularly as World Urbanization Prospects. He retired from the United Nations in November 2011.

Shobhakar Dhakal is the Executive Director of the Global Carbon Project, hosted at the National Institute for Environmental Studies, Tsukuba, Japan. With a background in engineering, economics, environmental management and policy analyses, his research focuses on various aspects of urban energy and environmental modeling, urban inventories and related analyses, including key urban sectors such as transportation with special emphasis on climate change issues. He has been working for over a decade on these topics and has been a key contributor to the several international assessments and studies. He currently serves as Co-Convening Lead Author of the chapter on human settlements for the Fifth Assessment Report of the IPCC.

David Fisk is a co-director of the BP Urban Energy Project, and holds the Laing O'Rourke Chair in Systems Engineering and Innovation at

Imperial College London. He was previously a holder of the Royal Academy of Engineering Chair in Engineering for Sustainable Development and Chief Scientist at the UK Department of Environment, where he was closely involved in the development of UK climate change policy. He is President of the Chartered Institute of Building Services Engineers, an Honorary Fellow of the Royal Institute of British Architects, and a Fellow of the Royal Academy of Engineering.

Arnulf Grubler is Acting Program Leader of the Transitions to New Technologies Program at the International Institute for Applied Systems Analysis, Laxenburg, Austria, and also Professor in the Field of Energy and Technology at the School of Forestry and Environmental Studies, Yale University, New Haven, CT (USA). Originally trained as a regional scientist, his research and teaching focuses on the long-term history and future scenarios of major technological transitions in energy, transport, and communication systems, with special emphasis on climate change, spatial patterns such as urbanization, as well as innovation as drivers of systems transformations.

Toshiaki Ichinose is an executive senior research scientist at the National Institute for Environmental Studies, Japan and also professor at the Department of Environmental Engineering and Architecture, Nagoya University. His academic background is in geography and civil engineering. His research interests include various aspects of urban environment systems, ranging from urban climatology, energy systems, to GIS applications in urban planning. He has performed many field surveys in China and other Asian countries and serves as a board member of the International Association for Urban Climate, and is also a member of WMO Expert Team on Training on Urban Climatology.

James E. Keirstead is a lecturer in the Department of Civil and Environmental Engineering, Imperial College London. His research focuses on the integrated modeling of urban energy systems and the links between urban form, consumer behavior, public policy, technical systems, and resource consumption. He is a chartered engineer and Member of the Energy Institute and a board member of the International Society for Industrial Ecology's Sustainable Urban Systems section.

Gerd Sammer is Professor and Department Head of the Institute for Transport Studies at the Vienna University of Natural Resources and Life Sciences, where he also served as Head of the Department of Landscape, Spatial and Infrastructure Sciences as well as Head of the Senate for Civil Engineering and Water Management. His academic background is in civil engineering, and his research focuses on transportation planning, transport surveys, and transport modeling at both the national as well as urban scales. He is also an accredited civil engineer.

David Satterthwaite is a senior fellow at the International Institute for Environment and Development, and editor of the international journal

Environment and Urbanization. He is also a visiting professor at University College London. A development planner by training with a doctorate in social policy, he works mainly on issues of urban poverty and the risks associated with it in regard to disaster risk and climate change. He contributed to the Intergovernmental Panel on Climate Change for the Third and Fourth Assessment Report (1998–2007), and is contributing to the Fifth Assessment Report that is currently under way. In 2004, he was awarded the Volvo Environment Prize.

Niels B. Schulz is a research scholar at the International Institute for Applied Systems Analysis, Laxenburg, Austria. His research focuses on indicators for society's material and energy metabolism for sustainable production and consumption, particularly at the urban scale. Previous positions include research analyst to the German Advisory Council on Global Change (WBGU), research fellow and team leader of the Urban Energy Systems Project at Imperial College London, and postdoctoral fellow at the United Nations University, Institute of Advanced Studies, in Yokohama, Japan. He holds a Ph.D. in ecology from Vienna University, where he researched changes in energy use and resource consumption during the industrial transformation of the UK. Past work also includes examinations of integrated measures for land-use and land-cover change such as human appropriation of net primary production, and ecological footprint analysis.

Nilay Shah is the Director of the Centre for Process Systems Engineering and co-director of the Porter Institute for Biorenewables and the Urban Energy Systems Project at Imperial College London. He is originally a chemical engineer by training, and his research interests include the application of multi-scale process modeling and mathematical/systems engineering techniques to analyze and optimize complex, spatially and temporally explicit low-carbon energy systems, including bioenergy/biorenewable systems, hydrogen infrastructures, carbon capture and storage systems and urban energy systems. Of particular interest is the question of how to integrate models which relate to different length and timescales. He is also interested in system design under uncertainty and in ways of characterizing risk.

Julia Steinberger is a lecturer in ecological economics at the Sustainability Research Institute of the University of Leeds (UK), and a senior researcher at the Institute of Social Ecology in Vienna, Austria. After obtaining her Ph.D. in physics at MIT (USA), she changed fields to focus on analyzing the drivers of energy and resource use at the international and urban levels, with the aim of identifying environmentally sustainable alternatives.

Alice Sverdlik is a Ph.D. student in urban planning at the University of California, Berkeley (USA). Her research interests include low-income housing, infrastructure, gender, and adaptation to climate change. Since 2009, she has worked at the International Institute for

Environment and Development, in London, and currently serves as a voluntary chapter scientist for the IPCC Fifth Assessment Report. She holds M.Sc. degrees in social policy from the London School of Economics, and in urban economic development from University College London, where she studied as a Marshall Scholar.

Helga Weisz is chair of the Department of Transdisciplinary Concepts & Methods at the Potsdam Institute for Climate Impact Research (PIK), Potsdam, Germany, and regular guest professor at St. Gallen University, Switzerland, and at Alpen-Adria University in Vienna, Austria. Originally trained in microbiology and cultural studies, her research and teaching focuses on material and energy flow analysis, industrial and urban metabolism, sustainable use of resources and society–nature coevolution.

1

Introduction and overview

Arnulf Grubler and **David Fisk**

1.1 Setting the urban stage

The decade 2000–2010 marked an important watershed in human history: for the first time more than 50 percent of the global population are urban dwellers. Estimates as reviewed in this book indicate that already some three-quarters of global energy use takes place in an urban context.

Given the robust trends toward further urbanization, the energy and sustainability challenges of equitable access to clean-energy services, of energy security, and of environmental compatibility at local through global scales cannot be addressed without explicit consideration of the specific opportunities and challenges of urban energy systems. The future development of the demand for energy cannot be described without understanding changes at the level of urban settlements. Research shows how the properties of urban areas across the world, while appearing diverse, are in fact scalable, revealing distinct patterns. Just as it is possible 'to fail to see the forest for the trees', it is possible 'to fail to see the city for the buildings'. A comprehensive energy assessment with an explicit urban focus and embracing a systems view has to date been missing. This book aims towards filling that gap. A systems perspective on urban energy use remains underdeveloped to date and paradoxes and conundrums in conventional analysis abound. A single urban agglomeration, such as greater Tokyo, generates more gross domestic product (GDP) than the venerable pioneer country of the Industrial Revolution – the current United Kingdom. And yet, our statistical reporting systems almost exclusively focus on nation states, as represented by Systems of National Accounts, Energy Balances, or similar international reporting standards. In fact, as detailed throughout this book, the difficulty of finding data at the urban scale starts with the search for an operational definition of urban areas and thus urban populations and their energy use.

From the perspectives outlined in this book the traditional territorial–administrative boundaries that define distinct 'cities' is complemented by an 'urban system' perspective, where the urban phenomenon is seen from a *functional* perspective that transcends traditional territorial or administrative system boundaries. Thus, an *urban energy system* comprises all components related to the use and provision of energy services associated with a functional urban system, irrespective of where the associated energy use and conversion are located in space, such as power plants linked to the urban area by a power grid and transport fuel requirements for movements of people and goods that are both local as well as global (arriving and departing at urban airports and ports). The full urban energy system entails both energy flows proper (fuels, 'direct' energy flows) as well as 'embodied' energy (e.g. energy used in the production of goods and provision of services *imported* into an urban system. Likewise direct energy uses in an urban system in turn become 'embodied' energy in the goods and services *exported* from an urban system). Such a functional perspective of urban energy systems highlights that urban locations and their growth (urbanization) are not only the clustering of people and economic activities in space, but also include the types of activities they pursue and the infrastructural and functional framing conditions (service functions) urban agglomerations provide. We contend that functional characteristics increasingly define urban areas and need to be better reflected in urban energy systems analysis, with a need to combine both 'production' and 'consumption' energy accounting perspectives. It is one of the (many) novel approaches of this book that the differences from alternative accounting methods are for the first time quantified, at least within the limits of the few available comprehensive urban energy accounts currently available. Likewise, this volume also extends – within the limits of available data – the traditional discussion of cities as defined by political and/or administrative boundaries towards urban agglomerations, including 'periurban' and larger metropolitan areas, all the way through to urban 'clusters' or 'corridors' to Doxiadis and Papaioannou's (1974) 'ecumenopolis'.

The future development of urban energy systems is characterized by specific challenges and opportunities. The high density of population, economic activities, and resulting energy use severely limit an obvious sustainable energy choice: renewable energy. In many modern larger cities locally collected renewables can provide for only some *1 percent* of urban energy use which implies large-scale *imports* of renewable energies generated elsewhere, much like in the currently dominating fossil energy systems. The diversity of activities and energy uses characteristic of urban systems opens numerous opportunities for intelligent energy management (e.g., electricity–heat cogeneration and 'heat cascading', in which different energy end uses can 'feed' on waste energy flows from energy conversion and industrial applications). Both *diversity and density* (at least above a critical threshold value of some 50–150 inhabitants/hectare of gross[1] settlement area) can be considered as key *strategic assets* of urban areas that help to use energy more efficiently by energy-systems integration, compact energy-efficient housing, and co-location of activities that can help to minimize transport distances and automobile dependence. The provision of transport services via high-quality urban public transport systems is a unique option for cities, generally not practical or economically viable in low-density sub-urban or rural contexts.

The vital urban infrastructures all depend on energy to function: water supply, treatment and waste water disposal, transport and communication systems, complex webs of food and material supplies and the resulting disposal of wastes, and, of course, energy supply itself for power and heating. Many urban infrastructures have shown great adaptability, but disasters such as hurricane Katrina's impact on New Orleans, or the 9/11 attacks on New York, show that urban systems and their populations are vulnerable because of infrastructural *interdependence*. Each urban infrastructure system is almost always managed in isolation, but this ignores their interdependence, and common vulnerabilities. Treating them in isolation also misses their potential synergies and efficiency gains. This highlights the importance of a systems perspective as well as of improved planning. But this will require new institutional frameworks and the inclusion of all relevant stakeholders to address the complex coordination issues across sectors and across spatial scales.

Urban agglomerations are dominant in terms of location of human activities, in the production and consumption of many goods and services, and in associated energy use. They are also unique centers of human capital, ingenuity and innovation, financial resources, and local decision-making processes, which are all 'human' resources/capital that can be mobilized, and at a vast scale. The urban scale is also the appropriate one to identify and realize many options in promoting energy efficiency that may not always be apparent at more centralized levels of policymaking.

1.2 Objectives and approach

Given the above, the broad objective of this book is to *assess* urban energy issues in an increasingly urbanizing world from a *systemic* perspective that focuses on the specific energy challenges and opportunities represented by urban settlements.

The specific objectives of this book are to perform first a global assessment to establish the order of magnitude of urban energy systems and their drivers, then develop some generalized explanations of urban energy use, and finally stress-test these explanations through case studies at the local and regional levels, drawing on specific examples of individual cities.

This book addresses the systemic and structural interlinkages of urban systems and how these interact within and outside traditional territorial urban system boundaries. It adds to the information and knowledge of traditional sectorial energy studies (buildings, industry, transport) by addressing the *integrated* issues specific and unique to urban systems. Sectorial perspectives are therefore addressed here only to the extent that they contain an explicit urban specificity, e.g. (public) urban transport systems, or urban energy cogeneration systems.

Urbanization is a multidimensional phenomenon that can be described from demographic, geographical, or economic perspectives. Empirical data are well developed for demographics (e.g. through regular population censuses) and for land-use perspectives (e.g. through remote sensing data). Conversely, there is a paucity of widely available and comparable economic data at the urban scale. Systems of National Accounts that underlie much of the available economic data were developed and continue to be used predominantly by national governments at the national scale, with comparatively few applications at the urban scale. The literature on urban land use and urban land-cover changes, while most valuable for describing a physical dimension of urbanization, is of limited use in an energy assessment despite the richness of quantitative data available. After all, it is not the square kilometers of urban extent that can explain urban energy use, but only the linkage of urban land use with demographic and economic data and characteristics as reflected through urban form and population density, infrastructure endowments, level and structure of economic activities, lifestyles of city dwellers, among others. Therefore, *demographics* (population sizes and their characteristics) is adopted quite naturally as a fundamental driver and core metric to discuss urbanization and urban energy use in this volume, drawing on the urban land-use change and economics literature only to the degree necessary to understand urban energy use and its variation through derived metrics centered on population, like population density, or per capita incomes and expenditures, in addition to more narrow disciplinary land use and economic perspectives metrics, such as urban extents/form or economic structure.

As comprehensive energy information and accounts at the urban scale are extremely limited, developing a robust assessment storyline from the bottom-up alone is challenging. Therefore in the analyses a mixed approach of both top-down and bottom-up perspectives is adopted, combining estimates derived from 'downscaling' or remote-sensing approaches with bottom-up statistical information where available.

1.3 Context and origins

The origins of this book arose from a collaborative effort by the authors of the various chapters in this book under the auspices of the Global Energy Assessment (GEA), often labeled 'the IPCC of energy' and released in 2012. In the context of this book, GEA assumes a special role, as it represents the first international assessment in which the issue of urbanization has been raised explicitly. The inherently interdisciplinary nature of any study of urbanization as well as the dispersed nature of knowledge and disciplines involved in urban studies perhaps explains why previous major international assessments refrained from addressing urbanization explicitly. (All authors of this book are encouraged in their scientific mission by the fact that the ongoing next assessment cycle of the IPCC includes a special chapter on 'human settlements' to which both the GEA assessment as well as this book and its authors are contributing.) Already early on in the GEA, the writing team of its urbanization chapter realized that there is much more material to cover in this first assessment of its kind that could be accommodated within the space limitations of the overall GEA. This book therefore presents a both deepened and enlarged assessment of energy in an urbanizing world than was possible within GEA. The editors and authors of this volume would like to express their deep appreciation of the initiative GEA has taken in addressing urbanization as a core energy topic and acknowledge with gratitude all the inputs and freely shared knowledge from all contributing authors to the GEA urbanization chapter[2] as well as from the wider scientific GEA community that led to this book volume.

1.4 A roadmap of this book

This book is divided into three parts. Part I The urbanization context (Chapters 2–5) provides the context of the urbanization phenomenon both from empirical and conceptual perspectives. Part II The urban challenges (Chapters 6–8) reviews major challenges ahead in moving urban systems towards more sustainable development paths. Finally, Part III Urban policy opportunities and responses (Chapter 9–12) addresses associated opportunities and policy responses in three key urban agenda items: transport, energy systems, and urban air-quality management. These core chapters are framed by an introduction (Chapter 1) that also includes a glossary of key terms, as well as by a summary and conclusion (Chapter 13).

Chapter 2 by Arnulf Grubler and Thomas Buettner provides a quantitative overview of the urbanization phenomenon. After reviewing the current state of urbanization, integrating multiple disciplinary perspectives including geography and land-use, demographics, economics, energy use, as well as other socio-economic and technological infrastructure indicators, the chapter focuses on the demographic dimension of urbanization, both from a historical as well as a futures (scenarios) perspectives. The chapter breaks new ground on several fronts, including a truly long-term historical perspective on urbanization that spans some 1,000 years, as well as an attempt to bracket the uncertainties surrounding future urbanization scenarios for the next 100 years that complement the long-standing central projections of the United Nations *World Urbanization Prospects.*

Chapter 3 by David Fisk provides a compact, 'tour de force' survey of the large-scale drivers and patterns of urbanization dynamics that adds the contextual and theoretical background to the quantitative assessment framed in Chapter 2. A key concept introduced in the chapter is that of *complexity* that is a hallmark of urban systems, both within (inter-urban) as well as in a global (intra-urban) context. An innovative feature of the chapter is to evoke Thorsten Hägerstrand's concept of 'timing space, and spacing time', with special reference to urban systems that are distinct in their time–geography 'choreography'. Maximizing 'contact potentials', urbanites tend to move fast, as evidenced in their significantly higher observed walking speed in urban spaces compared to rural settings. The chapter concludes with a synoptic discussion of the specificities of urban energy systems, emphasizing in particular the potentials for innovation-inducement effects of urban energy and environmental constraints that give reason for cautious optimism: urban systems are highly innovative and could pave the way to a much needed sustainability transition.

Chapter 4, by Julia Steinberger and Helga Weisz addresses the intricate issue of urban systems boundaries and their implications for urban energy and greenhouse gases (GHG) accounting. Addressing the issue of where to appropriately draw systems boundaries is of critical importance in inherently open urban systems; the chapter also provides a most useful connection to concepts developed within the framework of industrial ecology and urban metabolism studies. The chapter is not only an essential primer on alternative methodologies and concepts of urban energy accounting that has not been available in any comparable form to date, but it also provides a first quantification and apples-to-apples comparison of different urban energy accounting frameworks. The fact that such a comparison was possible only for two cities (London and Melbourne) illustrates best the long way still to go to arrive at widely agreed and standardized urban energy and GHG reporting frameworks.

Chapter 5 by Arnulf Grubler and Niels Schulz provides a combined 'top-down' and 'bottom-up' assessment of current urban energy use focusing on (direct) final energy use as most parsimonious and most widely available urban energy use indicator. The global, 'top-down' assessment of urban energy use estimates combines a range of spatially explicit data sets and estimates to provide a multiple-perspective quantitative overview of urbanization, which in its scope has not been available in the literature to date. A companion, 'bottom-up' assessment of urban energy use presents a novel and unique data set compiled for some 200 cities worldwide which is subsequently analyzed in more detail to examine the influence of macro-drivers on urban energy use, providing an empirical foundation for the review of urban energy use drivers literature (summarized in Chapter 9).

Part II opens with a sequence of three chapters reviewing urban sustainability challenges, the first of which is Chapter 6 by David Satterthwaite and Alice Sverdlik. The authors provide a comprehensive overview of the energy access and housing challenges for the urban poor, with an emphasis on the rapidly growing urban areas in low-income countries. Next to housing, also a range of energy services including cooking, lighting, as well as transportation are reviewed highlighting both challenges as well as innovative and successful policy responses for addressing urban poverty. Perhaps one of the most important key messages emerging and particularly valuable in an energy assessment is the importance of political and institutional settings and frameworks that can either further or block more energy-centered pro-poor policy initiatives in urban areas of low-income countries.

Chapter 7 by Niels Schulz, Arnulf Grubler, and Toshiaki Ichinose reviews a unique aspect (and constraint) of urban energy systems: the vast concentration of people, economic activities, and associated energy use constitute formidable density constraints in terms of energy demand, air pollution, and energy use proper, which at the scale and density of large urban areas gives rise to its own environmental constraint: the urban heat island effect. A particularly novel aspect of the chapter is the quantification of urban energy demand densities, which tend to be orders of magnitude above renewable energy flows that can be realistically harvested in urban areas. The magnitude mismatch between urban energy demand and renewable supply densities is all too often ignored in global, regional, or national-level energy and GHG reduction analysis and modeling highlighting the importance of the explicit consideration of urban scale energy systems.

Chapter 8 by David Fisk first provides a synoptic overview of energy supply constraints in an urban context including issues of energy security and reliability and opens the discussion of urban opportunities and response options to urban energy challenges, reviewed in more detail in Part III of this book (Chapters 9–12 below). The chapter reviews briefly the origins of sustainability concepts at an urban scale and then examines examples of proposed zero-carbon urban designs which are challenging both in their daring designs as well as in their investment requirements. The chapter then provides a generic overview of the main policy players and instruments at the urban scale that need to be leveraged to address the urban sustainability challenges.

Part III opens with Chapter 9 by Xuemei Bai, Shobhakar Dhakal, Julia Steinberger and Helga Weisz, which provides a detailed literature review of the drivers of urban energy demand. Not all of the determining drivers are amenable to policy making at an urban scale, but the comprehensive survey provides a most valuable context and a powerful illustration of path dependency in urban development paths that both illustrate opportunities for policy making in early development stages as well as significant constraints in more mature stages of urban development. A particularly novel feature of the chapter is the discussion of urban exergy analysis that provides a powerful tool in identifying opportunities for energy-efficiency improvements at an urban scale which may not be necessarily revealed through conventional energy analysis.

Chapter 10 by Gerd Sammer reviews urban transport trends and options in detail. Again the topic of path dependency emerges with density and public and non-motorized transport policies emerging as main framing conditions that can either perpetuate automobile dependence or aim at steering away from it. The chapter is also a powerful illustration of the value of a systemic view of urban transport as well as of the potential pitfalls of narrow policy approaches that often trigger 'vicious' development cycles, when ignoring fundamental relationships between urban transport infrastructure supply and transportation demand. A transition to more sustainable urban transport systems is feasible, but will require a wide range of well-coordinated policy initiatives that involve urban density, form, and usage mix, pro-non-motorized and public transport policies and infrastructures as well as better internalization of the external costs in terms of accidents, congestion, and environmental impacts associated with overdependence on automobility in urban areas.

Chapter 11 by James Keirstead and Nilay Shah discusses both urban energy planning and systems design, as well as integrating energy-sector policies into a larger framework. The chapter draws on a novel modeling approach ('SynCity') that combines both traditional energy systems optimization frameworks with actor-based and spatially explicit modeling frameworks. With the help of 'controlled experiments' the chapter illustrates the respective impacts of policy interventions along three main opportunity areas: urban form and density (and their influence on transport energy demand), buildings efficiency, and urban energy systems integration (cogeneration) and optimization for an illustrative, hypothetical city. The simulations quantify the respective orders of magnitude of policy interventions at the urban scale (see also Chapters 5 and 9): urban form/density and buildings efficiency have a comparable order of magnitude potential impact each on lowering urban energy use. Conversely, the potential for urban-scale energy (supply) systems integration and optimization is, while significant, much smaller. Energy demand thus takes precedence over energy supply management, with the largest efficiency gains evidently possible by a combination of both demand and supply-side measures.

Chapter 12 by Shobhakar Dhakal concludes Part III of this book, with a review of urban air pollution, focusing on major trends, particularly in the rapidly growing urban areas in Asia, as well as in reviewing a range of success stories in controlling traditional air pollutants at the urban scale. While the magnitude of the challenge remains enormous, particularly in view of the anticipated urban growth in Asia, Africa, and the Middle East, there are also many encouraging signs of rapid adoption of low-emissions standards and technologies aiming at 'bending' long-term emission trends 'downwards'. As also evoked in other chapters throughout this book, institutional capacity (or lack thereof) remains a key area of concern for many cities, especially at smaller scales.

Lastly, Chapter 13 by Arnulf Grubler and David Fisk provides a summary and conclusion to the collective assessment effort documented in this book. Major trends, challenges, and opportunities in a rapidly urbanizing world are summarized and policy directions highlighted. While generic conclusions are possible at this stage of our assessment of how to energize our future cities, important knowledge gaps remain. Any generic conclusions will also have to be interpreted in view of the high degree of local context specificity of urban policy and decision making which requires to complement the global analysis presented here by local and place-based assessments as well.

1.5 Glossary of major terms and concepts used in this book

Energy accounting

primary energy comprises all energy forms as extracted from nature (e.g., crude oil) within a system (nation, city) or imported from outside the system boundary under consideration (e.g., gasoline or biofuel imports to a city).

total primary energy supply (TPES) is a quantity defined in the statistics of the International Energy Agency (IEA) and comprises all the primary energy extracted nationally (and by extension also regionally, including urban areas), as well as the net balance between imports and exports of primary and secondary energy. This includes commercial and noncommercial energy, such as traditional biomass or combustion of waste products for energy use.

secondary energy is energy that is exchanged and/or transformed within the energy sector (and situated between the primary and final energy levels) and a frequent accounting unit of urban energy studies. Examples include (refined) fuel oil that is input to oil-fired electricity generation, or natural gas that fuels a district-heating plant.

final energy is the energy that is 'consumed'[3] by the end-user (e.g., purchased motor fuels or electricity for household, industry, or service-sector energy consumers). Usually, it has undergone several stages of transformation, transportation, and final distribution.

upstream energy is the primary or secondary energy required to produce the final energy used within an urban area. The energy level (i.e., primary or secondary energy) that is referred to in upstream energy calculations should be made explicit in energy reporting.

embodied energy is the primary energy required to produce and transport the goods and services imported and exported to and from an urban area (often referred to as 'indirect' or 'gray' energy in the literature) as opposed to direct energy flows (final, secondary, or primary energy) used directly as fuels in an urban area. For comprehensive urban-energy accounting, imported embodied energy should be added to accounts for direct urban energy use, whereas the embodied energy of exports from an urban area to its regional, national, or international hinterland should be subtracted. (An analytical shortcut that avoids the intricacies of urban import/export energy accounting is represented by the 'consumption based' accounting approach. This ignores urban productive activities and instead estimates the direct (including upstream) and embodied energy use related to the consumption of goods and services in an urban area. This approach provides valuable information on the importance of consumption choices in diets, housing, or personal transport, but it misses important opportunities for urban energy policy making in energy systems and in manufacturing and service industries.)

Energy versus greenhouse gas (GHG) accounting

Whenever possible the focus of this volume is on urban *energy* to maximize comparability. However, in many cases the available literature needs to be reflected. This often focuses on CO_2 or total GHG emissions, without, however, always specifying the underlying energy data and the sources (energy, agriculture, land-use change, etc.) and GHG species included in the analysis. While this in some cases reduces comparability across different literature sources, it provides a valuable additional perspective that reflects the available literature.

production- or consumption-based accounting describe energy accounts that focus on alternative system boundaries to define urban energy use:

- The 'production approach' usually defines the system from a 'top-down' energy-production, conversion, and distribution perspective typical of the energy supply sector (including utilities). It typically accounts only for direct energy flows, either consumed as final energy by households, industry, services, and the public sectors or used as secondary and primary energy in the required 'upstream' energy transformations that deliver final energy. It is the urban equivalent of traditional national energy balances, but with the national geographic system boundary replaced by an urban one. Hence, this approach is also referred to as 'territorial' accounting, as the system boundary is defined primarily through geographic and/or administrative spatial boundaries. Its advantages arise from more readily available data and the direct information value for local decision making. Its main disadvantage is the omission of embodied energy flows, as urban areas are characteristically net importers of resources and energy to a larger extent than that of typical national economies. This accounting approach is the energy equivalent of the GHG accounting and inventory frameworks that underlie the reporting of GHG emissions (e.g., the Organisation for Economic Co-operation and Development (OECD)/ Intergovernmental Panel on Climate Change (IPCC) emission inventory guidelines).
- Conversely, a 'consumption approach' takes a 'bottom-up' perspective of final consumption (typically measured through expenditures) and estimates all associated direct and embodied energy or GHG flows of the bundle of goods and/or services represented by the various expenditure categories based on proxy energy (or GHGs) per US$ value indicators derived from lifecycle assessments (LCAs) or (typically) national Input–Output (I–O) tables. Its main advantage is a more comprehensive system boundary, which gives a more realistic picture of the energy (or ecological) footprint of an urban area that imports from national and global hinterlands. Its disadvantages include its data intensiveness (and hence paucity of published accounts) and unresolved methodological intricacies. For example, the representativeness of national I–O tables for specific urban areas characterized by different economic structures and price levels, data-quality issues for estimates of energy embodied in international trade flows, and the accounting of energy use associated with urban infrastructure investments (e.g., road construction) that are not included in household-expenditure statistics as constituting a public good are all important methodological issues that await resolution. Also, accounting for the energy demand for transport (within and between the system boundaries under consideration) represents a particular challenge

in this approach. The difference in the two accounting approaches becomes particularly pronounced in trade-intensive open urban economies and when considering industrial versus service-sector oriented urban economies. In urban areas with an important industrial and manufacturing base (with associated significant exports), a consumption-based accounting approach generally leads to lower energy use (and GHG) estimates compared to a production-based accounting approach. Conversely, for service-oriented urban centers that typically import all energy and energy-intensive materials and goods, a consumption-based accounting approach yields substantially higher energy use (and GHG) numbers compared to production-based energy accounting.

Urban system boundaries

urban system boundaries a conceptual boundary in a physical or functional space which encloses all components that form the urban system.

spatial and administrative definitions of 'urban' No set of criteria for defining urban areas or urban populations is accepted and applied universally, and each country uses its own set of criteria which are subsequently aggregated by the United Nations (UN) to derive global and regional estimates of urban populations and of urbanization levels. Among the 228 entities (countries and territories) recognized in the UN 2007 World Urbanization Prospects (WUP) statistics, 83 use administrative criteria, one uses economic criteria, 57 define minimum population sizes of settlements as thresholds for their urban classification, four use 'urban characteristics' as criteria and 48 apply any combination of the above criteria. In six entities the entire population is considered urban, three do not have any urban population and in 26 cases the definitions are either unclear or not provided. Nonetheless, the UN aggregates of national statistics are broadly globally consistent with a uniform urban definition as settlements with more than 50,000 inhabitants and a minimum population density of 150 people/km^2 (World Bank, World Development Report 2009; Uchida and Nelson 2008).

urban (and urban systems) as used herein refers to the entirety of different entities that define urban along either administrative, political, functional, geographic, or socio-economic criteria.

towns are in size typically larger than a village, but smaller than cities. They often do not have the functional diversity (of industrial, residential, and service-sector activities) and also often have a different, subsidiary, legal status to cities.

cities are localities defined according to legal and/or political boundaries and an administratively recognized urban status that is usually characterized by some form of local government that may well extend beyond the city's immediate boundary.

urban agglomerations are defined through the contours of a contiguous territory inhabited at urban density levels without regard to administrative boundaries. These usually incorporate the population in a city or town plus that in the areas lying outside of but adjacent to the city boundaries. Agglomerations can be monocentric (in the case of a central city and its surrounding suburbs or satellite cities) or polycentric (containing several city centers) and enclose parts of adjacent periurban (semiurban) or rural areas.

metropolitan areas include both the contiguous territory inhabited at urban levels of residential density and additional surrounding areas of lower settlement density that are also under the direct influence of the city (e.g., through frequent exchanges, commuting, etc.). Several metropolitan areas can fuse together to form urban 'conurbations' or urban 'corridors'.

functional definitions of 'urban' The definitions of urban systems based on common denominators of socioeconomic activities as revealed through the exchange of economic goods, people, or information.

commuter sheds (or commuter-fields) are analogies to watersheds, and are defined by 'travel to work areas' in which populations are grouped according to the most likely direction of commuting for the purpose of employment, typically using travel surveys of daily work commutes. The idea is based on a gravitational concept of cities as 'attractors' and is often used to illustrate that the functional system boundary of what comprises an urban area is much larger than traditional administrative and/or geographic boundaries of a city. Definitional and measurement issues are far from trivial in this concept.

Urbanization indicators

level of urbanization is a stock measure of either the total amount of a variable (urban population, urban land area, urban GDP) or as its share to the total entity (e.g., proportion urban, the percentage of the total national population that is classified as urban).

urbanization rate measures the change of the proportion (share) of urban population over time. Also often referred to as urbanization (process).

Additional concepts

Zipf distribution or rank–size rule is a mathematical distribution function of cities, typically examining population size (or other indicators like urban areas, infrastructure endowments, etc.) as a function of ordinal rank, and is generally associated with the work of George Kingsley Zipf. There is a mathematical relationship between the size of the largest (ordinal rank = 1) to the second largest, all the way down to the nth largest city. The observed stability of the rank–size distribution implies that, in the aggregate, cities tend to grow at a uniform rate (even if individual cities grow at different rates that yield a change in their respective position in the overall stable rank–size distribution). The implications of this distribution for urban energy systems is the need to analyze cities across the entire range of the distribution curve.

Zahavi 'rule' refers to a concept of urban transport postulated by Yacov Zahavi, in which transport choices are conceptualized as maximizing distances traveled subject to time budget (approximately one hour per day) and money budget (approximately 15 percent of disposable household income) constraints.

Notes

1   Residential housing plus commercial/industrial zones plus areas for technical and recreational infrastructures including roads, public transit, and parks and other open spaces.

2   Gilbert Ahamer (University of Graz, Austria), Timothy Baynes (Commonwealth Scientific and Industrial Research Organisation, Australia), Daniel Curtis (Oxford University Centre for the Environment, UK), Michael Doherty (Commonwealth Scientific and Industrial Research Organisation, Australia), Nick Eyre (Oxford University Centre for the Environment, UK), Yunichi Fujino (National Institute for Environmental Studies, Japan), Keisuke Hanaki (University of Tokyo, Japan), Mikiko Kainuma (National Institute for Environmental Studies, Japan), Shinji Kaneko (Hiroshima University, Japan), Manfred Lenzen (University of Sydney, Australia), Jacqui Meyers (Commonwealth Scientific and Industrial Research Organisation, Australia), Hitomi Nakanishi (Commonwealth Scientific and Industrial Research Organisation, Australia), Victoria Novikova (Oxford University Centre for the Environment, UK), Krishnan S. Rajan (International Institute of Information Technology, India), Seongwon Seo (Commonwealth Scientific and Industrial Research Organisation, Australia), Ram M. Shrestha (Asian Institute of Technology, Thailand), P.R. Shukla (Indian Institute of Management, India).

3   The colloquial term 'energy consumption' is in fact at odds with the first law of thermodynamics: energy can neither be created nor destroyed, but only transformed (converted).

Part I

THE URBANIZATION CONTEXT

2

Urbanization past and future

Arnulf Grubler and **Thomas Buettner**

2.1 Introduction

As discussed in the following chapters, urban systems are distinctly different with respect to their energy systems and levels and patterns of energy use and hence merit special consideration in energy analysis and policy. This chapter therefore aims to provide a wider context of the urbanization phenomenon. We first outline the multiple disciplinary perspectives on urbanization, before proceeding to a discussion of the urbanization phenomenon from a dynamic perspective, discussing both historical as well as future scenario trends. Given available data, the discussion of past and future urbanization trends focuses on the demographic dimension. However, as discussed in greater detail in Chapters 5 and 9 below, such a demographic perspective on urbanization needs to be complemented by considerations of the economic and geographical dimensions of urbanization as well in order to be able to explore the energy implications of urbanization.

2.2 The multiple dimensions of urbanization

The process of urbanization involves multiple dimensions, characterized by different theories, methodologies, and literatures that follow four distinct disciplinary perspectives: demography, geography, economics, and sociology.

The resulting effects of the scale and concentration of human activity in urbanized areas are the focus of research in various disciplines. Economists often emphasize the benefits of scale of larger labor markets in cities and agglomeration effects of clustering of various industrial and service activities (Krugman 1991; Fujita et al. 1999; UNIDO 2009; World Bank 2009). Climatologists discuss the consequence of urbanization on albedo changes, radiation balances, and weather patterns (Kalnay and Cai 2003; IAUC and WMO 2006; Souch and Grimmond 2006). Transport planners are concerned to avoid negative externalities of urban density, such as traffic congestion. Environmental researchers study typical patterns of the generation and distribution of pollutants in urban centers and the exposure of target populations to such hazards (McGranahan et al. 2001; McGranahan and Marcotullio 2007). Social scientists are investigating particular urban social structures and challenges, and urban cultural modes of creativity and innovation that result from the immediate proximity of many million people that can exchange, cooperate and profit from high degrees of specialization. Understanding the consequences of urbanization on energy use in general, however, is an area of research that has attained surprisingly little attention in empirical studies, given the relevance of urban areas for overall energy demand (IEA 2008), their particular vulnerability to energy supply disruption, and their potential for energy savings and climate-change mitigation.

The economic, geographic, and sociological perspectives of urbanization are discussed in greater detail as driving forces of urban energy demand (see Chapter 9). Following these different disciplinary perspectives, four complementary concepts describe the process of urbanization:

1. The demographer's approach emphasizes population. To a demographer, urbanization is the process by which a rural population becomes urban, either by people migrating from rural to urban places or by a transformation of a rural settlement into an urban settlement. That is, the population is becoming urbanized reflected by an increase in the proportion of the population classified as urban dwellers.
2. In the geographer's approach, a defined geographic area gradually loses the characteristics associated with rural areas (e.g., dominance of agricultural land uses) and gains characteristics associated with urban areas (e.g., built-up land, and high density of buildings and technological infrastructures such as mains drainage) – the region is becoming urbanized.
3. The economist's approach focuses on the process of economic structural change away from primary economic activities (agriculture, forestry, mining, etc.) toward manufacturing and services (secondary and tertiary economic activities) where typically capital increasingly substitutes for labor. This usually involves spatial concentration and co-location of economic actors[1] that profit from agglomeration externalities: the economy is becoming urbanized.
4. In the sociologist's approach individuals (and family units) move from rural to urban areas, and take on urban characteristics: individuals and their aggregate (i.e., society) are becoming urbanized.

The demographic study of urbanization is among the oldest research traditions and also the most quantitative, including scenario projections into the future. Along with economics it is also the dimension with the closest direct causal link to urban energy use. This is the reason why in the subsequent discussion the demographic perspective of urbanization is highlighted. The economic, geographical, and sociological perspectives of urbanization in turn are discussed in greater detail as driving forces of urban energy demand (Chapter 9).

Table 2.1 illustrates these multiple dimensions of urbanization at the global level for the year 2000, the latest common year for which the

Table 2.1 Various indicators of urbanization for the year 2000 at the global level in absolute amounts, and as percentage urban with associated uncertainty ranges derived from the literature or estimated in this volume

Indicator		Value	Range	References for Uncertainty Range
Area	(1000 km²)	2929	313–3524	Schneider et al., 2009
	% of total	**2.2**	0.2–2.7	range of GlobCover-Grump data
Population	(million)	2855	2650–3150	Uchida and Nelson, 2008
	% of total	**47**	44–52	Size threshold: 50,000–100,000
GDP (MER 2005$)	(billion)	32008		this assessment
	% of total	**81**		not available
Final energy use	(EJ)	239	176–246	this assessment
	% of total	**76**	56–78	(see Chapter 5)
Light luminosity	(million NLIS)	33		
	% of total	**57**	50–82	this assessment
Internet routers	(number in 1000)	8592		
	% of total	**96**	73–97	this assessment

Notes: MER: Market Exchange Rates; NLIS: Light Luminosity Intensity Sum (Index)

Sources: Crovella (2007); Grubler et al. (2007); NOAA (2008); UN DESA (2010)

various indicator data sets are available. With the exception of remotely sensed urban land areas (that turn out to display the largest uncertainty range of a factor of ten) and population for which data are available directly at the urban scale, all other indicators represent derived estimates, combining spatially explicit data sets of urban extents consistent with UN urban population estimates with other spatially explicit data sets of human activity available. The urbanization indicators adopt a necessarily narrow definition of urban activities as those taking place within the spatial urban boundaries consistent with the available spatially explicit socio-economic data sets. For instance, the final energy use estimates refer to the direct use of final energy (fuels and electricity) in urban areas and excludes upstream energy conversion losses (e.g. in electricity generation) and energy embodied in raw material and goods imported into urban areas[2] for which spatially explicit data sets or estimates are unavailable. Adopting a multi-dimensional perspective on urbanization reveals that in terms of economic activity, energy use, and technological infrastructures the world is already predominantly urban today and is bound to yet further urbanize in the future.

2.3 Urbanization in a historical context

Cities have existed for millennia. Cities and their associated urban agglomerations have and continue to play key roles as centers of government, production, and trade, as well as knowledge, innovation, and productivity growth (UN 2008). Yet during most of the past, the number of city dwellers remained exceedingly small – the majority of the population continued to live in rural areas and undertook manual labor in agriculture.

Before 1800 large cities (by contemporary standards) were exceedingly rare, with Chang'an and Baghdad likely the first examples of cities that approached one million inhabitants (Chandler and Fox 1974; Chandler 1987). This started to change dramatically with the (rapid) improvement of agricultural labor productivity, which displaced people to new industrial jobs and migration into cities (the Industrial Revolution). It needs to be noted that historically, the economic structural shift away from agricultural employment significantly preceded the corresponding shift towards urban residence, albeit the development lags became shorter over time (Figure 2.1). Whereas in the nineteenth century the economic and demographic structural shifts unfolded as lagged time trends,[3] they more or less occurred simultaneously in the twentieth century, highlighting a new qualitative dimension of the urbanization phenomenon.

As a result of these structural shifts, many cities grew to much larger sizes, a growth that was enabled by new transportation infrastructures (in particular railways, later on also road infrastructures); rural settlements were transformed into cities, and new cities were founded. The rate of urban population growth (the twin result of natural growth in urban populations plus net in-migration into cities) exceeded significantly that of the overall population and resulted in a secular trend toward *urbanization* (i.e., an increase in the rate or percentage of a population living in urban areas).

There are no comprehensive global data sets available to enable to describe urban population growth and urbanization before 1950, the year systematic UN statistical data series start for all countries and the world total. However, the formidable statistics compiled by Tertius Chandler (1987) reveal a striking regularity in the global rank–size

Figure 2.1 Ratio of agricultural to non-agricultural workforce (solid lines) and rural to urban population (dashed lines) for England and Wales, United States, Japan, and Brazil (logarithmic scale). Empirical data have been approximated by a set of coupled logistic equations that appear as straight lines in the logarithmic transformation shown
Source: Adapted from Grubler (1994)

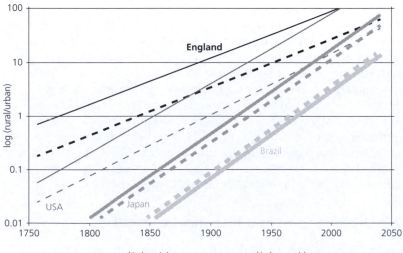

distribution of cities (Figure 2.2) which can be used in inferring[4] global urban populations before the year 1950 (Figure 2.3).

Prior to the Industrial Revolution (i.e. pre-1700), the urban population of the world did not exceed an estimated total of some 30 million, or between 4 to 6 percent of the total population. The stable and comparatively low extent of urbanization rates worldwide illustrates the limitations of agricultural productivity and resulting limited agricultural surplus production (for feeding urban populations) in the pre-industrial era. Our estimates suggest that urbanization trends accelerated only after 1700 with urbanization rates increasing gradually to between 10 to 18 percent in the period 1900 to 1930, with total urban population ranging between 200 to 400 million worldwide in the first three decades of the twentieth century. Because of their lead in industrialization and agricultural productivity growth, the more developed countries had an urbanization rate of some 30 percent by the 1920s, and developing countries of only some 6 percent (UN DESA 1973).

It is also interesting to note that, historically, the geographical locus of urbanization has been in Asia, the continent with the largest population. This Asian dominance of urbanization, gave way to a European (Western Europe and its offshoots in North America and Australia) interlude in the nineteenth and twentieth century. An over-proportional share of the world's largest cities in this period were and are located in Europe and its offshoots as a consequence of accelerated urbanization in the regions of early and pervasive industrialization. This "European century" of world urbanization (and industrialization) is however giving way to a globalized urbanization phenomenon in the twenty-first century in which both the largest number of urban settlements as well as the largest urban agglomerations will be increasingly outside the traditional industrialized countries.

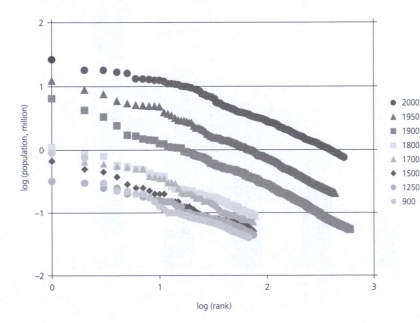

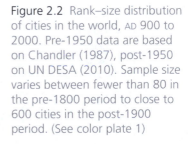

Figure 2.2 Rank–size distribution of cities in the world, AD 900 to 2000. Pre-1950 data are based on Chandler (1987), post-1950 on UN DESA (2010). Sample size varies between fewer than 80 in the pre-1800 period to close to 600 cities in the post-1900 period. (See color plate 1)

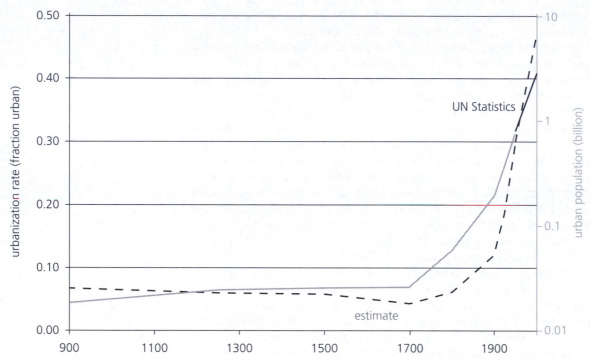

Figure 2.3 Estimates of global urban population (billion, right axis, solid green lines) and fraction urban (fraction, left axis, dashed red lines) AD 900–1950 inferred from Chandler (1987), and UN DESA (2010) data 1950 to 2000 (black lines). (See color plate 2)

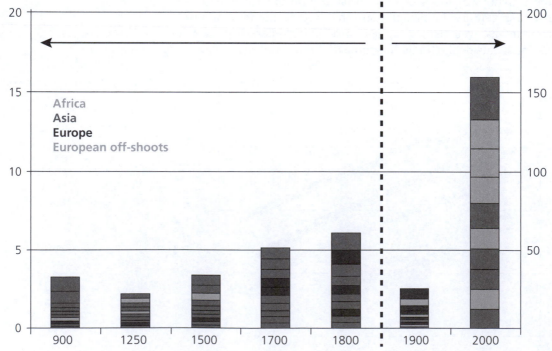

Figure 2.4 Urban population of the 10 largest cities, by continent, in millions, AD 900 to 1800 (left axis) and 1900 to 2000 (right axis). (See color plate 3)
Sources: data: Chandler (1987); UN DESA (2010)

The UN reconstruction of global population trends from 1950 onwards (UN 2008, UN DESA 2010) documents a historically unprecedented global population growth in the second half of the twentieth century, peaking at about 2 percent annually between 1965 and 1970. Global population growth rates have declined since and current projections indicate a leveling off of global population growth toward the third quarter of the twenty-first century.

In 1950, the world population totaled some 2.5 billion. In 55 years, the world's population grew by about 4 billion people to approximately 6.5 billion in 2005. A defining feature of this growth is its diversity across regions, especially between more- and less-developed ones. Having entered the demographic transition[5] earlier, more developed OECD countries and countries undergoing economic reform (Eastern Europe and the former USSR) increased their population between 1950 and 2005 by a comparatively modest 62 percent and 51 percent, respectively. All other regions, by contrast, more than doubled their populations. The fastest population growth occurred in the Middle Eastern and African regions, where the population more than quadrupled between 1950 and 2005. The population in Latin America and the Caribbean more than tripled, and in Asia, already the most populous region of the world, in 2005 the population was 2.7 times larger than that in 1950.

The inevitable degree of uncertainty in pinning down urban population and urbanization levels does not affect an assessment of the dynamics of the urbanization process, provided the system of national definitional criteria does not change too often over time. Occasional definitional changes of individual countries do not markedly affect global or regional trends.

While populations in all regions grew over the past half century, urban populations grew even faster (Table 2.2). Between 1950 and 2005, the UN DESA (2010) estimate suggests that the Middle Eastern and African region multiplied its urban population more than ten-fold, increasing from 44 million to 478 million. Asia registered in 2005 an urban population almost six times higher than it had in 1950 (175 million versus 1.3 billion), and the urban population of Latin America and the Caribbean increased more than five times, from 68 million to about 429 million people (UN DESA 2010). By 2005 about half of the global population (3.2 out of 6.5 billion) was urban. By the time of the publication of this volume (2012), more than half of global population will be urban.

As the urban population grew, also the relative weight of this population segment increased. In 1950, just about 30 percent of the world's population of 2.5 billion people lived in urban areas, while the vast majority still lived in rural areas. Even for the more advanced OECD region, in 1950 57 percent of its population lived in urban areas, while a substantial proportion of its population (43 percent) lived in rural areas. All other major regions were predominantly rural. Latin America and the Caribbean, and the countries currently undergoing economic reform (Eastern Europe and former USSR) already had high urbanization levels (41 percent and 38 percent, respectively) in

Table 2.2 Urban total population (millions) and as a percentage of world total urban population, and total population (in millions) for the five world regions and the world since 1950. Countries undergoing economic reform = Eastern Europe and former USSR, OECD90 = OECD member countries as of 1990

GEA Region	Urban Population				Total Population	
	1950	**2005**	**1950**	**2005**	**1950**	**2005**
	in Millions		in percent of world		in Millions	
OECD90	340	730	46.6	23.1	593	963
Countries undergoing Econ. Reform	102	254	14.0	8.0	269	404
Asia	175	1,276	24.0	40.3	1,237	3,478
Middle East and Africa	44	478	6.0	15.1	266	1,115
Latin America and the Caribbean	68	429	9.4	13.5	165	552
World	729	3,167	100	100	2,529	6,512

Source: UN DESA (2010)

1950. Populations in the Asian and the Middle Eastern and African regions were still predominately rural, with just about 14 percent and 16 percent of their populations living in urban areas, respectively.

By 2005, the world had become almost 50 percent urban, but with significant differences across the regions. In the OECD and Latin America and the Caribbean countries, three out of four people lived in urban settlements; and in the Asian and the Middle Eastern and African regions about two out of five people were urban dwellers.

2.4 Urban futures

Trends in population aging and urbanization are inevitably important to the immediate and medium-term demographic future. Population growth is still a persistent element of demographic trends in developing countries. For some developed countries, however, population growth has not only tapered off, but may be reversed with population decline as a new defining demographic feature.

Significant population decline is currently expected to persist for countries undergoing economic reform (i.e. the countries of the former Soviet Union/USSR and of Eastern Europe), shrinking by about 39 million people from 404 million in 2005 to 365 million in 2050. All other regions are projected to increase their populations between 2005 and 2050 (UN DESA 2010). There is, however, great diversity not only between regions, but also, and more significantly, within the regions and between countries. There are also different causes for the continued population growth within most regions. Populations in the OECD regions, which have experienced low or even very low fertility rates during recent decades, are expected to increase slightly, mainly because of net gains from international migration. For example, Western Europe is projected to lose about 1.4 million people between 2005 and 2050, when net migration gains are included. In the absence of that gain, in 2050 Western Europe would have about 15 million fewer inhabitants than in 2005 (UN DESA 2010).

Population growth in the other regions is caused by very different factors: momentum that stems from young populations arising from survival through early childhood, and above-replacement fertility, the latter increasingly less important because of a continued fertility decline. Combined, these factors generate the addition of 1.1 billion people to Asia, 1.2 billion to the Middle Eastern and African region, and 173 million to Latin America and the Caribbean to 2050.

Against this background, the growth of the global urban population is estimated to range from some additional 2.7 to 3.2 billion people, depending on the projected urbanization rate of growth (see Box 2.1). The generally larger growth of the projected urban population not only means that urban settlements will absorb all of the population growth between 2005 and 2050, but also that there will be a sizable redistribution of rural populations to urban areas. In addition, urban growth will be predominantly a phenomenon of the less-developed regions.

Figure 2.5 summarizes the GEA scenario results on the global level, and Figure 2.6 summarizes five world regions in terms of urban and rural populations. The continued growth of the world's urban population is set against a different growth path of the rural population.

Globally, rural population growth is projected to come to a halt around 2020, when it will reach a peak at about 3.5 billion people. This figure is unaffected by the uncertainty of the urbanization scenarios and is a major, robust conclusion. The finding also adds urgency to corresponding efforts to improve energy access for the rural poor because if energy does not reach them soon, they will have an additional incentive to seek access in urban areas. After entering a path of negative growth, the global rural population is projected to range from 2.8 to 3.3 billion by 2050 and to decline even further thereafter. However, the global picture masks stark regional differences: the Middle Eastern and African region will likely experience rural population growth until 2050 (or at latest to 2080 in GEA-L), while for all other regions the beginning of the decline in rural populations is imminent or is already occurring (in the OECD90 region and in countries undergoing economic reform).

In 2050, the world is projected to be 70 percent urban with a comparatively narrow scenario uncertainty range of 64–70 percent. Latin America and the Caribbean, as well as the OECD countries, are expected to approach 90 percent urban, about the level of urbanization of the United Kingdom or Australia today.[6] By 2100, world urbanization levels could range from 71 percent to 89 percent with a corresponding urban population from 6.7 to 8.4 billion people. Thus, even in the lowest scenario, the world urban population in 2050 will be larger than the entire global population today.

Box 2.1 Urbanization projections methodology

1 The UN World Urbanization Prospects (UN WUP)

The UN WUP provides biannual estimates and projections of core demographic indicators of urbanization for all 229 countries or areas of the world and for the most populous urban settlements. The data include time series of total populations by urban and rural residence for the period 1950 to 2050, and of populations in 590 large urban settlements with 750,000 and more inhabitants for the period 1950 to 2025. In total, the entire WUP database contains estimates and projections for 4,501 urban locations, covering two-thirds of the global urban population.

The WUP use a projection method that is described and discussed extensively (UN DESA 1971, 1974, 1980; Ledent 1980, 1982, 1986; Rogers 1982; O'Neill et al. 2001; National Research Council 2003; Bocquier 2005; O'Neill and Scherbov 2006). Such long-term projections include many uncertainties, reflected in the three urbanization scenarios discussed here. The method uses the growth difference between urban and rural populations or between city populations and total urban populations to model the dynamics of the urbanization process, an approach that is equivalent to a logistic curve fit (UN DESA 1974). Starting with the most recent observed growth rates, future trends are calculated by taking into account the empirical observations that the pace of urbanization slows down as the proportion of the urban population increases (a characteristic feature of logistic growth processes beyond the 50 percent level of its asymptote). Consequently, the observed growth differentials are adjusted such that they approach a hypothetical norm obtained from past empirical observations (UN DESA 1980). A distinguishing characteristic of the WUP is that only one central projection is performed in which the ultimate upper level of the urbanization process is not constrained; that is, all countries could ultimately converge to an urbanization rate of up to 100 percent.

2 Other urbanization projections

There are no real independent urbanization projections to the UN Urbanization Prospects as alternative scenarios invariably use the UN historical and current data as model inputs and also deploy a comparable methodological framework. The literature, however, reports scenario variants on the UN projection by either extending the time horizon beyond 2050 (the end-year reported by the UN projections) and/or alternatively relaxing the convergence assumption toward a 100 percent urbanization rate. For an OECD study on the climate vulnerability of port cities, Nicholls et al. (2008) extend the UN projections to 2100 and then apply a constant fraction to port cities at the national level to determine future port-city populations exposed to climate-change risk. In an integrated scenario exercise that explores future uncertainty in GHG emissions, Grubler et al. (2007) also extend the UN urbanization projection to 2100 and develop two additional scenario variants in which the asymptotic urbanization levels are varied to explore the implications of lower urbanization. These three urbanization-rate scenarios were then combined with three alternative total population-growth scenarios (low, medium, and high) to determine the uncertainty range of future urban populations. This alternative scenario method estimated the uncertainty in the level of world urban population to range from 4.7 to10.5 billion people by 2100, and from 5.6 to 7 billion compared to the 6.3 billion projected by the latest UN Urbanization Prospects projection for 2050 (UN DESA 2010).

3 GEA urbanization scenarios

As part of the Global Energy Assessment (GEA) a range of energy sustainability transition scenarios were developed. Methods and qualitative and quantitative assumptions that underlie the GEA scenarios are described in detail in Riahi et al. (in press). In a collaborative effort, also the issue of urbanization scenarios was explored. In this context, two main features of the GEA scenarios need to be emphasized (1) their focus on energy issues and (2) their normative nature; that is, their objective to describe alternative, but successful, pathways toward an energy sustainability transition. As a result, the scenarios are based on a single central demographic–economic development scenario (e.g., based on the most recent UN medium population projection). The scenarios also describe pathways in which current widespread economic and energy disadvantages of poor rural populations are addressed successfully, for example with universal energy access achieved by ca. 2030. As a result, the qualitative scenario storylines describe developments in which, arguably, the pressure for rural to urban migration is significantly relieved, which suggests their reflection in the GEA urbanization scenarios. Therefore, three urbanization scenario variants based on a single medium population projection were developed. One scenario (GEA-H) uses the UN Urbanization Prospects projection directly as input, whereas the other (lower) scenarios (GEA-M and GEA-L) adopt the methodology outlined in Grubler et al. (2007) using a model recalibrated to the most recent (2010) UN urbanization data. The resulting scenario differences in projected rural and urban populations bracket the potential impact of successful rural development policies that relieve urban migration pressures (GEA-L) compared to largely unaltered patterns in rural and/or urban locational advantages (GEA-H) and illustrate to policymakers the potential effects of altered policies that change the locational advantage of rural versus urban places.

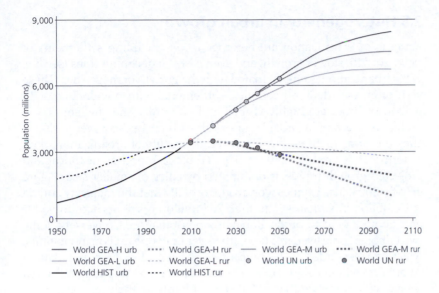

Figure 2.5 Global urban (solid lines) and rural (dashed lines) population (in millions). Historical trends 1950–2005 and scenarios to 2100 combining a medium population growth scenario with three alternative scenarios of urbanization rate growth (based on Grubler et al. (2007)) and comparison to the most recent UN Urbanization Prospects projection (circles).
Source: Adapted from UN DESA (2010)

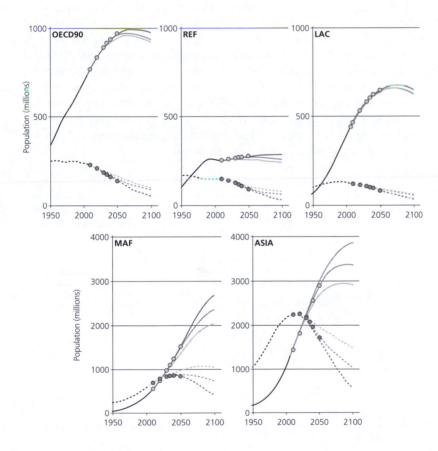

Figure 2.6 Regional scenarios of urban and rural population (millions). Historical trends 1950–2005 and scenarios to 2100 for five world regions (in millions). Regions: OECD90, OECD countries as per 1990 membership; REF: Countries undergoing economic reform, i.e. the reforming economies of Eastern Europe and the ex-USSR; LAC: Latin America and the Caribbean; MAF: Middle East and Africa; ASIA: developing economies of Asia. Urban population, solid lines; rural population, dashed lines. UN Projections (circles) and three alternative scenarios developed for the Global Energy Assessment (GEA).

2.5 Heterogeneity in urban growth

Patterns of urbanization are heterogeneous, including settlements of rapid growth, slower growth, and even cases of declining cities (see Box 2.2). Urbanization is often equated with the growth of megacities. These vast, often crowded and complex, settlements with populations of 10 million or more are highly visible, and epitomize the challenges and problems of a rapidly urbanizing world, with pervasive poverty, slums, stressed infrastructures, etc. However, the reality of urbanization, both in terms of current settlement sizes (Table 2.3) and as historical and projected growth trends, is dominated by cities of smaller size (Figure 2.7). By 2005, just 19 cities worldwide formally met the megacity criteria (>10 million inhabitants) and their 268 million residents accounted for only about 8 percent of the global urban population. Conversely, some two billion people lived in cities with fewer than one million inhabitants, which corresponded to 61 percent of the world's urban population. About one billion people or 34 percent of the urban population lived in smaller cities of fewer than 100,000 inhabitants[7] in the year 2005.

The virtual absence of smaller cities in statistical reporting, data collection, and modeling poses a serious problem and adds substantial uncertainty to any statement on urbanization trends. However, this should not lead to the erroneous conclusion not to focus analytical and policy attention on those cities that dominate the global urbanization phenomenon, most with <100,000 inhabitants. These numerous smaller cities pose a triple challenge for policymaking: data are largely absent, local resources to tackle problems with urban growth may be limited, and governance and institutional capacities to implement policies for more sustainable urban growth can be weak.

Table 2.3 Population in urban locations by city size class in 2005

Size class	City population	Proportion of total urban population
		2005
	Millions	**Percent**
Total urban population	3,167	100
<100,000	1,069	34
100,000–1,000,000	932	29
1,000,000–5,000,000	673	21
5,000,000–10,000,000	209	7
>10,000,000	284	9

Source: UN DESA (2010)

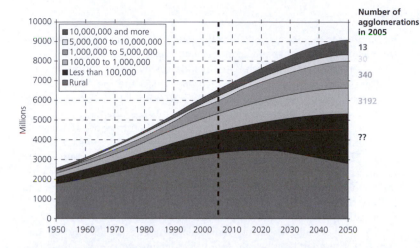

Figure 2.7 Population by residence and settlement type (millions). Historical (1950–2005) and projection data (to 2050) (for 2005 statistics, see Table 2.3)
Source: adapted from UN DESA (2010). (See color plate 4)

Box 2.2 Shrinking cities

The growth of an urban settlement is not a law of nature, as cities have stagnated in size over long periods of time, or even shrunk, and may well do so in the future. In addition, as national and international production and trade patterns change, some cities may lose their economic or strategic advantage and so shrink or even be abandoned entirely, for which history also provides ample examples. In the United States and Europe, many of the great nineteenth and early twentieth-century steel, textile, and mining centers and ports have lost economic importance and population (Cunningham-Sabot and Fol 2009; Wiechmann 2009); so too have some of the major manufacturing cities – for instance Detroit as a center of motor vehicle production. Also, in various high-income nations, from the 1970s onwards there appeared to be a reversal of long-established urbanization trends nationally or within some regions with a net migration from large to small urban centers or from urban to rural areas. This was labeled counter-urbanization, although much of it is more accurately described as de-metropolitanization because it was population shifts from large/dense metropolitan centers or central cities to smaller urban centers or suburbs or commuter communities.

Few systematic analyses have been performed on contemporary shrinking cities (UN HABITAT 2008). Based on the latest assessment by the UN (UN DESA 2010), of the 3,552 cities with 100,000 and more inhabitants in 2005, 392 experienced a combined population loss of 10.4 million people between 1990 and 2005.[ᵃ] While this seems a rather modest figure, given the 3.2 billion urban dwellers in 2005, for some cities population decline was substantial: a few hundred thousand inhabitants. The decline in urban population can, in some cases, be far from gradual but very fast, as in the eastern German city of Hoyerswerda (Pearce 2010):

> In the 1980s, it had a population of 75,000 and the highest birth rate in East Germany. Today, the town's population has halved. It has gone from being Germany's fastest-growing town to its fastest-shrinking one. The biggest age groups are in their 60s and 70s, and the town's former birth clinic is an old people's home. Its population pyramid is upturned – more like a mushroom cloud.

Fearing decay, vandalism, and costs, high-rise apartment buildings are now being torn down.

Cities also shrink as a reaction to an often painful economic restructuring process, as observed in countries of Eastern Europe and the former USSR undergoing economic reform. Most OECD90 countries have had persistently low fertility, well below replacement levels, for extended periods of time. If low fertility rates continue, overall population decline will become a reality. Shrinking cities amid a growing or stable national population could then be replaced by a regime of national and city populations shrinking simultaneously. How this will affect urbanization remains one of the biggest challenges for understanding long-term urbanization trends.

The energy implications also remain an important area to be explored. Shrinking urban populations yield reductions in total urban energy use. Such trends are, however, likely to be offset to some extent by the effects of potentially larger residential floor space available for the remaining population as well as increases in transport energy use associated with lower population density. The economic viability of urban transport and energy infrastructures in shrinking cities may also be challenged, but on this aspect no data or studies are available currently.

Notes

1   In economics this is referred to as agglomeration externalities that arise from economies of scale and economies of scope effects.

2   The urban final energy use estimates include however the energy embodied in goods and services produced in urban areas and exported (i.e. not consumed within the urban geographical boundary). For a detailed discussion of urban energy accounting concepts, see Chapter 3.

3   A widely used model in the technological change literature used to describe structural shifts is based on the logistic substitution model (Fisher and Pry 1971). The S-shaped patterns of structural change can be conveniently linearized via the log (F/(1–F) transformation, where F denotes the fractional share of the indicator under consideration (agricultural vs. total jobs, or rural vs. total population in this example).

4   By assuming that the rank–size distribution established for the largest world cities equally holds for smaller settlement sizes (as current data suggest), an integral can be calculated yielding an estimate of global urban population and a corresponding urbanization rate (proportion of population that is urban) when compared to the total estimated global population (based on Deevey 1960).

5   The term demographic transition refers to a state change in the two fundamental demographic drivers: fertility and mortality. The demographic transition involves a phase change from a demographic condition characterized by high mortality and fertility rates, to a state characterized by both low fertility and mortality rates. The transition from high to low mortality rates generally precedes a similar transition in fertility rates by several decades, a lagged development that translates into a period of accelerated population growth.

6   Projected urban populations by 2050 at the country level (UN DESA 2010) suggest that the two countries with the by far largest urban populations by 2050 will be China and India with some 1 and 0.9 billion, respectively. Ten countries are projected to have urban populations above 100 million by 2050 and, with the exception of the US (370 million), all of these are currently developing countries (projections by 2050 in millions: China, 1040 million; India, 875; Nigeria, 220; Brazil, 200; Pakistan, 200; Indonesia, 190; Bangladesh, 125; Mexico, 110; Philippines, 100).

7   Calculated mainly as a residual to the estimates of total urban population.

8   A UN HABITAT (2008) report suggested that in a sample of cities analyzed that comprised 1.9 billion inhabitants (some 60 percent of the world's urban population in 2005), an estimated 400 million people lived in cities that experienced population declines in the 1990–2000 period (260 million in high-income countries, 140 million in low- and middle-income countries). However, this statement is misleading, as it ignores the comparatively modest decline in absolute population numbers of these cities: 10 million (UN DESA 2010) to 13 million (UN HABITAT 2008) between 1990 and 2000. The apparent decline in many city populations is also rather a population shift from central city areas to suburbs or other urban areas.

3
Urbanization dynamics

David Fisk

3.1 Introduction

How do urban centers grow and change over the long timescales of energy infrastructure? If patterns exist, they would have consequences for investment in energy systems and transport. Modern towns and cities are generally recognized in the urban studies' literature as formally complex, largely self-organizing systems (Allen 1997; Amaral and Ottino 2004; Batty 2005) that exhibit important agglomeration effects (National Research Council 2003; World Bank 2009). 'Formally complex' is meant here in the sense that predicting detailed behavior would require an inconceivably vast information requirement. They are 'self-organizing' in that interactions between partially informed citizens and between citizens and the city's 'hardware' form self-reinforcing patterns of land use and allocation of time. What was prescriptive 'town planning' in the early twentieth century is now seen more realistically as the facilitation of spatial patterns trying to emerge naturally from the complexity (Hall 2002). So-called 'econophysics' represents a discipline that has sought to use the data we now have available stretching longitudinally over centuries, and cross-cutting around the earth to extract a common high-level narrative.

The long life of the urban settlement demonstrates that individual urban settlements can show extraordinary resilience to external forces. Long-standing settlements have usually had to reinvent their *raison d'être* many, many times. Large cities usually confront physical constraints to existing paradigms first and have to develop innovative solutions (such as freshwater-supply networks, the use of clean secondary fuels and the use of subway systems) that are then taken up by innovation diffusion to smaller settlements. Each city sits within a wider economic or social fabric and so in understanding urban dynamics it is important to distinguish between production and consumption viewpoints. Thus final energy *demand* constraints often serve as a shaping factor in settlements. For example urban settlements, at least in times of peace, are more easily and cheaply served by transport when located in a valley, on a river, or on the seacoast. To overcome the disadvantage of location elsewhere would demand more energy consumption. The location of intensive energy use has its own effect on urban layout. Differentiation of rental value often takes place downwind of the pollution of energy-intensive industries – 'wrong side of the tracks'. This applies equally to 'passive' sources of energy. Prevalent solar and wind conditions often account for orientation patterns of streets and squares in traditional urban complexes. In contrast, energy *supply* considerations, the consumption perspective, have taken a secondary role in determining urban *form*. Indeed, the reverse has often been true – urban form creates a demand that selects from the energy mix often bringing into being demand for new secondary energy products. For example large nineteenth-century settlements needed the high-energy density of coal because of the transportation problems associated with carting larger volumes of wood into the dense center. Later, the cities needed the coal processed into town gas to provide better-quality services, such as street lighting. Urban settlements moved into inhospitable environments once they could buy high-grade energy to run mechanical and electrical building services (Banham 1969). Over the long perspective of future energy-infrastructure planning and population growth, the infrastructure implications of self-organizing properties of urban space need to be taken into account. A historical rarity would be a reconfiguration of the urban form to reflect constraints on forms of energy supply. One exception might be cities where power supply is erratic, as in some cities in the developing world where urban infrastructure has outrun investment in power. Such cities tend to be low rise even in the central district because no one wants to be trapped in a lift in the daily blackout!

3.1.1 Inter-urban complexity

Econophysics provides some fascinating insights into the interactions between urban settlements. Some of the most compelling evidence that complexity theory is at work is that groups of settlements (i.e. within the migration range of a mobile population) as ranked by population size approximately conform to Zipf's rule – size is inversely proportional to rank (Zipf 1949; see Figure 3.1). As earlier discussions have highlighted we should not expect too much precision in meeting this rule, but it is 'true' at least to within the inevitable indeterminacy of the definition of an urban population. Within such a data set each urban center is in a dynamic social and economic equilibrium with the remaining national urban system and, although urban settlements may often think of themselves as 'self-contained', they are far from it. The implication is that at the margin, individual urban settlements show no *total* economies or diseconomies of scale. If they showed either property, they would simply disperse to a small uniform size or accumulate into a single megatropolis. This equilibrium does not need to apply to any specific economic activity (e.g., the location of energy-intensive industries or finance centers). Large cities and small towns do different things better and worse, but large and small appear to show no special advantages in the economy seen as a whole. Individual cities and towns in the modern world are part of a single urban system. The implications of the urban rank–size rule for this assessment suggest that any undue overemphasis in energy planning on the top end of the distribution curve (i.e., an exclusive focus on 'megacities') is certainly not warranted, as smaller cities constitute the majority of the continuum of the rank–size distribution curve.

At present, few data or studies are available that explore the implications of the rank–size rule for urban energy use. Bettencourt et al. (2007) identified economies-of-scale effects for urban energy infrastructures for gasoline distribution and electricity distribution grids (cables); i.e. large cities (in terms of population size) have less than proportionally larger energy infrastructures than smaller cities, at least in Germany and the United States. In terms of energy use the sparse available evidence is mixed: transport energy use (gasoline sales) seems to be somewhat lower in larger cities in the United States, which appears plausible in view of the impacts of higher population density on lower automobile dependency. Conversely, electricity use appears to grow somewhat over-proportionally with city size (United States, Germany, and China). However, it is unclear whether this is a genuine urban scale effect or simply reflects fuel substitution effects where larger and denser cities exhibit higher preferences for clean, grid-dependent energy carriers (more electricity and/or gas and less oil and/or coal) compared to smaller cities. This remains an important area for future urban energy research.

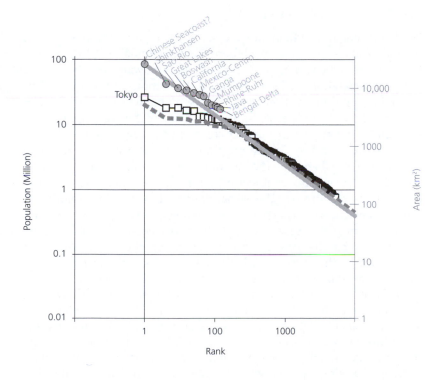

Figure 3.1 Rank–size distribution of global cities. Note the irregularity in the distribution at the largest cities when defined by their administrative boundaries. Conversely, by aggregating individual cities to 'urban clusters' or 'corridors' the regularity of the global city rank–size distribution is maintained. The designation of an emerging urban cluster along the Chinese seacoast is entirely speculative. The currently largest urban cluster is found in Japan along the Shinkansen corridor, with some 70 million inhabitants. The rank–size distribution of global cities not only extends to their populations (square for cities and circles for urban agglomerations, left axis), but also to their area (land cover; dashed line, right axis)
Sources: Figure 2.2 and Marchetti (1994); city area distribution from Yamagata (2010), based on Kinoshita et al. (2008)

Zipf's law was once thought suggestive of some deep underlying mechanism. It appears in many other contexts. It now seems likely that the (approximate) power-law size distribution results from many coexisting stochastic processes with a similar statistical effect to that of a 'long tail' power law (Sornette and Cont 1997). The only constraint on candidates for one of the underlying processes is that the pattern of stochastic growth has to retain a long tail power-law distribution on average as the national population grows. In fact, very simple growth processes can generate these power laws. For example, among the very simplest, Ijiri and Simon (1975) obtained a power-law tail just by assuming that each year's population growth is randomly spread across settlements with a probability in proportion to the settlement size. The key to maintaining a power-law distribution that is robust over time is not the modeling of the largest settlement, but the model's mechanisms for creating and retiring the smallest.

The robustness of the Zipf law means mechanisms that populate the large number of smaller settlements play a more important role in urbanization dynamics in most countries than the growth of the few largest cities, as amply confirmed by urbanization statistics (see Chapter 2). It adds to the conclusion herein on the urgent need to study smaller-sized ('normal') cities in terms of their energy and environmental implications.

3.1.2 Intra-urban complexity

While the smallest settlement is, to some extent, a matter of statistical definition, at its most basic level there is an economic and physical limit to the size of settlement that can manage its own water, sewage, power, and administration collectively, at a few thousand people. This reasoning replicates within the urban settlement itself (Batty 2008). Quanta of local areas arise similar in size to the minimum urban settlement that appear in the early history of a large city. On the city map they are marked out by physically distinct districts. There is a subtle interplay between the quantum size suggested by technical constraints of urban engineering and the economic and social expectations of citizens. The social construct of the 'neighborhood' must embed the physical constraints that ensure viable shared service and connectivity. It is thus possible to think of a city as a set of adjoining cells that can have perceived social and economic characteristics, in part determined by the cells with which they join. Their changes with time can then be rather as if they were cellular automatons that can be flipped in status according to rules dependent on the status of adjoining districts as well as through internal causes that looked stochastic as seen from outside. This dynamic model hints that above some rate-of-event threshold of cell-switching states (say between 'up and coming', or 'going down hill'), the city remains a heterogeneous map, with districts changing their various designations from time to time but without districts of a similar type ever actually coalescing into a much larger similar aggregation. Below some coexistence threshold between differing types of district, groups of districts start to cluster together in one class and the designations segregate spatially (Schelling 1969), which can act as a precursor to the formation of slums and ghettos as development capital is diverted to more attractive areas. This 'phase change' can be seeded by an infrastructure change (an investment in a noisy airport or a large public space or a waste incinerator). Cellular automata models can reproduce many of the key social spatial statistics of urban settlements (Batty 2005). Energy infrastructure development, such as district heating, thus has to take into account not only developments in overall urban energy demand, but local variations, some of which may represent permanent changes from the conditions that attained at the beginning of the project.

The attachment of new zones and transformation of others reflect urban-area attempts to gain the advantages of agglomeration while internally separating discordant functions (Fujita et al. 1999). This game is intrinsic to land use changes. Zoning, whether imposed or emergent, is then a natural property of an urban settlement. The dynamics of these processes have long been recognized as complex (Forrester 1969). Indeed, modern urban planning frequently uses techniques from statistical mechanics to find the most likely aggregate properties from the nearly random effect of the thousands of citizen choices (Wilson 2000). Administrative boundaries do not always follow the expansion of an urban area and care needs to be taken, as noted repeatedly elsewhere in this book, as to whether an urban settlement's quoted energy statistics refer to the current political or to the practical spatial reality.

An urban system can be seen as a process with 'land' as an input, in effect making the town or city a 'space machine' (Hillier 1999) reorganizing land use and connections to use land to best effect. Whatever the actual process, only a limited number of stable connectivity patterns seem to emerge for expanding urban spaces. This may, in part, reflect that when expansion is contemplated, the pre-existing connection patterns that will have to be adapted to include the new zone have a very strong influence on the optimal connections *within* a new zone or district. Such an assumption underlies much of the modeling using cellular automata, applied to understanding patterns of urban growth. Preferred attachment rules for new entrants are a common property in large complex systems that grow while remaining dynamically stable (Fisk and Kerhervéa 2006).

One constraint on the connectivity network is that expansion needs to occur such that the whole urban settlement remains socially and economically viable at all stages of growth. It may, in part, explain why the world's urban areas adopt one of only a few generic configurations. The 'idealized' European city physically expands by absorbing smaller satellite centers and creating annular transport networks, to form a scale-free connectivity that enables access on foot to public transport. Low-income, low-rent areas cluster around those areas where transport to work is cheap.[1] In contrast, the 'idealized' North American grid city expands along its major axes and maintains a uniformity that avoids overstressing its original center, again using beltways to relieve congestion at intersections from clipping journeys. Although grid cities are superficially uniform, partitioning of rental value is often triggered by environmental factors (downwind, downriver, across the tracks) and agglomeration by local economies of scale. Both these urban configurations were created historically by capital investment in the city that roughly matched the influx of population. Conversely, in many developing countries the migration rates from rural areas currently run ahead of capital formation, creating serious issues of energy access and infrastructure development (see Chapter 6) and a third model of a peri-urban development.

3.1.3 Time and the city

While urban geography is naturally expressed in terms of space, an urban settlement might equally be thought of as a complex time machine. Each citizen has a '24 hour budget' and this needs to be juggled to leave time to get somewhere to take time to do something with others. Since the urban rationale is to bring specified people together at the same place at the same time to get things done, motion within the area needs to meet socially determined constraints in absolute time. Zahavi's rule (Zahavi 1974) notes the rough constancy, in many countries over many decades, of the time spent traveling during a working day. This presumably reflects the influence of many shared social norms, such as when workplaces start and finish business, when meals are taken collectively and when shared entertainment begins and 'latecomers are not admitted'. Since different modes of travel involve different amounts of physical work by the traveler, it has even been suggested that the social norm also sets upper limits on how tired the traveler can be on arrival, which gives a gloss to the feasible means of travel mode (Kölbl and Helbing 2003).The adjustment of land-use patterns then 'solves' the consequent set of time constraints. Thinking of a city as a vast set of service connections self-consistently feasible within given absolute time windows is a complementary dynamics paradigm to that of the space machine with some fascinating explicative power.

Transport improvements (or the prospect of them) tend to induce land-use changes and attract investment, and do not in *the long run* save traveling time (as reviewed in Metz 2008), despite this being the initial rationale of the infrastructure development. Faster travel times lead to larger high-rise central business districts or suburban sprawl, not more time in bed. In urban areas the local density of travelers on the move directly affects the average speed of travel they can each achieve and also indirectly decides safe maximum speed. A city with high values of speed averaged over complete journeys is supporting journeys of longer length within social norms, and so individual service points (e.g. shopping centers) sweep larger areas of clients. Increased average speed supports larger agglomerations (e.g., large retail shopping malls are a consequence of an urban freeway system) and so gains economies of scale. Higher speeds can give greater separation between incompatible land uses without breaking the social norms that schedule journeys.

As a consequence of land-use adjustments, high-speed cities have longer average journey lengths, but this means their growth dynamics can relax to lower densities with greater local accumulation. It would seem that hour or so traveling time represents the effective limit to the radius of commercial activity of an urban settlement. The resulting interdependency of *transport*-energy consumption from higher speeds on urban density is roughly linear when plotted against the mean interpersonal distance extracted from the urban density per unit area (Kenworthy and Newman 1990). This is the kind of relationship that might be expected from a time of travel dominated by traveler– interaction effects. Since the total start-to-finish journey time is the key determinant of location, there are densities below which only individual motorized transport can provide a plausible service within a plausible time. By the same token, as urban density increases the delays from the interaction between travelers becomes increasingly significant and energy consumption ceases to fall as energy is wasted in congestion and traffic queues. Very high-density centers offer little advantage to transportation because congestion is so severe (especially true in cities in which both the mass transit and private transit systems become embroiled in the same gridlock). Urban areas in this state tend to spill outwards and create a self-organizing critical density consistent with the underlying social norms.

3.1.4 Economic complexity

The new insights of complex system 'econophysics' when applied to urban settlements complement rather than displace traditional urban economics. The timescales and spatial statistics employed tend to be of a different character. Econophysics focuses on changes over long timescales and widespread spatial averages and so often matches the timescales of energy infrastructure investment. Traditional economics models the shorter-term dynamics that manage the scarcity and surplus of individual space. The overlap is the understanding of the dynamics of rents, both in the form of local taxation and 'ground' rents. The interplay of rent and infrastructure is another under-researched area, despite its vital role in shaping the urban fabric. These influence income distribution and the dynamics of investments and also change the economics of urban- versus rural-based activities (Irwin et al. 2009). Planning constraints (such as requirements to connect to a district heating scheme or meet certain energy performance targets) become adjusted into the value of land, much as the value of land reflects other positional advantages and disadvantages. Differences between urban settlement dynamics in different societies may then reflect more profound institutional differences (e.g., as reviewed by Diamond 2004) than simply choice of technology.

3.1.5 Future dynamics

The energy demand and energy mix of modern urban settlements normally follows rather than leads urban expansion (Seto and Shepherd

2009). However, historical demographic generalities about urban growth hide the continual innovation within the settlements themselves. For example new transport technologies that made such a fundamental contribution to urban form in the nineteenth century were not spontaneous innovations but arose from solving a problem caused by a pre-existing horse-powered technology. So projecting the urban future on the basis of past trends is dangerous if attention is not paid to changes in detailed mechanisms of innovation and adaption that underlay those earlier trends. Indeed, there are already signs that indicate that this new urbanized century will have to be different in character from the urban–rural mix that preceded it.

For example, the earlier simplicity of Zipf's law is now retained for the world's largest cities only if the associated metropolitan region is treated with the city as one (see Figure 3.1 above). If airline connectivity is a new paradigm of interurban transport, the connectivity of the largest cities appears not to have caught up with expectations from a natural extension of the power law to describe airline connectivity of smaller towns (Guimerà et al. 2005). Indeed, scaling arguments (Kühnert et al. 2006) suggest that the frequency with which settlements need to find paradigm-shifting innovations must increase as they grow. Since breakthroughs are, by definition, unforeseeable, there is a future scenario in which the growth of the world's largest metropolitan areas simply falters because they have yet to find how to think their way out of their current problem.

Urbanization forms part of the demographic transition and is expected to be associated with a stabilization and then decline in world population around the middle of this century. As a consequence some urban settlements may face retractions in some unspecified form. How built form relaxes to new configurations (such as urban farming in derelict industrial [smoke stack] districts) poses a fascinating question to a new understanding of urban structure that can no longer be based on past experience of patterns of expansion and development.

3.2 Specifics of urban energy systems

In principle, urban systems are not fundamentally different from other energy systems in that they need both to satisfy a suite of energy-service demands and to mobilize a portfolio of technological options and resources. However, urban systems also have distinguishing characteristics that set them apart:

- A high density of population, activities, and the resulting energy use and pollution (see Chapter 7 for a more detailed discussion).
- A high degree of openness in terms of exchanges of flows of information, people, and resources, including energy.
- A high concentration of economic and human capital resources that can be mobilized to institute innovation and transitional change.

Urban areas are characterized by high spatial densities of energy use, which correspond to their high population concentrations, and by low levels of energy production and extraction: cities are loci of resource management, processing, trade, and use, rather than of resource extraction or energy generation. All settlements depend on a hinterland of agriculture, forestry, mining, and drilling; in the present fossil-fuel economy, this hinterland has a global reach. Indeed, the same may be true of many rural areas, which, in developed countries, often focus on a few crops. These specialized rural areas also require imports of energy, goods, and services, which often may be on the same scale, per capita, as those of urban areas. Indeed, spatial division of labor is a characteristic of modern societies, with consequences for local energy use and policies. In industrialized countries both urban and rural areas depend heavily on energy-intensive industry, which may be located in or outside cities. If heavy industry is located outside urban areas, urban energy consumption may apparently be lower than rural energy consumption, even though urban dwellers also consume the products from industrial activities.

High levels of energy demand open possibilities to reap significant *economies-of-scale* effects in energy systems, in supply as well as in transport and distribution. (It is not a coincidence that, historically, many major, large hydropower resources were developed to supply electricity to large urban agglomerations, from Niagara Falls in the Unites States to Iguazu (Itaipu) in Brazil.) Simultaneous to cities being loci of a wide diversity of activities, significant *economies of scope* are possible. The wide range of different energy applications from high-temperature industrial processes down to low-temperature residential home heating, and even to the energy provision of greenhouses, allows the maximization of energy efficiency through better source–sink matching of energy flows in the system, be it through conventional cogeneration schemes or through more complex waste-heat 'cascading'. These can, however, only be realized if diverse energy uses in a city are sufficiently mixed and co-located to allow these concepts of 'industrial ecology' to be implemented. On the negative side of density, typical urban agglomeration externalities are important: low transport efficiencies through congestion and high pollution densities add urgency to environmental improvement measures. Retrofitting a high-density built environment can also incur many transaction costs that would not apply in low-density developments. However, it is no coincidence that as a result cities have always been the first innovation centers for environmental improvements (Tarr 2001; Tarr 2005).

The latter perspective is perhaps the most fundamental for the transformation into more sustainable energy systems. Urban agglomerations are *the* major centers and hubs for technological and social innovation (Kühnert et al. 2006), as they both dispose of and mobilize formidable resources in terms of human, innovation, and financial capital. Bringing these transformations to fruition may ultimately be of greater long-term environmental significance than any short-term environmental policies. So, energy and environmental policies in an urban context may have substantial leverage in inducing further much-needed innovation in the core, where such activities take place.

Note

1   In cities of developing countries, economic pressures to relocate slums that enjoy a transport locational advantage are therefore high. Conversely, locating the poor far outside the city with poor (and expensive) transport access (as was the case in apartheid South Africa) further disadvantages them socially and economically.

4

City walls and urban hinterlands: the importance of system boundaries

Julia Steinberger and **Helga Weisz**

4.1 Introduction

Measuring the energy use of cities from readily available data is no easy task, with comparisons compounded by the absence of widely agreed measurement concepts and data-reporting formats. And yet for reasons of scientific enquiry, policy guidance, and political negotiation, the important issue of 'attribution' needs to be addressed. The seemingly easy answer to the question – how large is the energy use (or associated GHG emissions) of a city and what can be done to reduce it? – can vary enormously as a function of the alternative geographic and functional system boundaries chosen. Therefore, this section reviews the different issues that must be addressed in urban energy assessments and aims to clarify the various concepts and definitions, and help make their differences more transparent. In a modification of an old adage that only what gets measured gets controlled, this chapter postulates that only what is measured correctly and transparently at an urban scale is useful for policy guidance.

As a simple example of the importance of boundaries in urban energy assessments consider the issue of the administrative/territorial boundary chosen for defining a given city. Barles (2009) studied the fossil-fuel use of Paris, its suburbs, and the larger Parisian metropolitan region. The per capita fossil-fuel consumption was lowest in the city of Paris, and increased as the region considered expanded. This phenomenon is likely caused by a combination of inherent and apparent factors: the inherent factor would be the lower transportation energy required by areas of higher population density (in central Paris with its formidable Metro system, compared to its suburbs); the apparent effect would be the changing of the system boundary to encompass more energy-intensive industrial activities located beyond the city center.

Different methods for assessing urban energy exist, and, ultimately, must be chosen on the basis of informing policy and research. Generally, urban-energy assessments must be oriented either to physical, and hence local, energy flows (a 'territorial' or 'production' perspective) or, if trade effects are included, follow economic exchanges linked to energy use (a 'consumption' perspective) (Ramaswami et al. 2011).[1] Much of this section will be devoted to the description of these different approaches and their contrasting results. The indeterminacy in defining a single value for 'urban energy' should not be misinterpreted as a flaw in the urban systems perspective. It reflects that the data are approaching the actual final-decision level at which the purpose of the decision, to some degree, can resolve many of the statistical and data ambiguities. For example, administrative boundaries and a production perspective are appropriate system boundaries if the decisions are to be undertaken by local administrations. Final energy-use data remain an essential and useful tool for analysis of energy efficiency and for crafting policies for improved efficiency. Conversely, a 'consumption' perspective on urban energy and GHG use helps to raise awareness that, ultimately, urban energy and GHG management cannot be relegated to an energy optimization task, but equally involve changing lifestyles and consumption patterns. The joint consideration of the production and consumption perspectives is most likely to yield a full assessment of urban energy,[2] albeit to date the literature and database for such comprehensive perspectives is extremely limited.

In assessing a variety of local and upstream contributions to the urban metabolism,[3] it is sometimes tempting to aggregate the disparate elements into one indicator. For instance, the ecological footprint[4] (see Rees and Wackernagel 1996) is increasingly used to describe the impact of urban resource use (e.g., in London, Barcelona, and Vancouver). The ecological footprint of an urban area is invariably far larger than the surface area of the city itself, by a factor of 120 and 180 for London and Vancouver, respectively. This leads to the facile (but erroneous) conclusion that the city is 'unsustainable' because it is not self-sufficient metabolically. In fact, cities are part of an exchange process, whereby they produce manufactured goods and services while depending on a hinterland for their supplies. Conversely, the economic existence of this hinterland is dependent on the demand from the city. Moreover, keeping particularly land-demanding and polluting-production activities, such as agriculture, mining, or heavy industries outside the city, while utilizing the potential of high-density agglomerations to support energy-efficient infrastructures and services, will likely have less overall environmental impact than a system of small-scale local autarky, as suggested by the ecological footprint.

Alternative indicators frequently discussed are the relative magnitude of resource use and emissions (per capita, household, or income), compared to rural or other urban populations, or environmental limits, such as carbon accumulation in the atmosphere, which is unsustainable because above a certain accumulation level large-scale regime shifts in the earth's climate system are likely to occur (Lenton et al. 2008). Since the ecological footprint is, in any case, driven by fossil-fuel carbon emissions, studying urban energy use directly appears to be a more constructive and policy-relevant approach.

4.2 Energy accounting methods

Table 4.1 classifies the various methods used to estimate urban energy use. This classification utilizes two main criteria: the basis of the data and the definition of energy users. The first two methods are based on physical flows, and produce 'territorial'- or 'production'-oriented energy balances; the other two focus on economic flows, and are either territorial or 'consumption' oriented. The economic based models (regional economic activity and economic I-O approaches) are further distinguished by the level of sectorial detail.

The 'final energy' method uses urban specific physical data, such as energy statistics from utilities or fuel sales, as the data basis. Users are defined as energetic end users within the city boundaries. Energetic end users are the 'consumers' of final energy (such as electricity, heat, gasoline, or heating fuels). By disaggregating the final energy use by sector, one can differentiate between residential, commercial, and industrial uses. It is important to note that the consumer of final energy in energy analysis (i.e. firms and households) is not the same as the

Table 4.1 Overview of urban energy-accounting frameworks

Approaches	Data basis	Definition of energy users	Position along energy chain	Upstream or embodied energy	Territorial/ Production or Consumption approach
Final energy	Physical	Final user (energetic)	Final	Not included, can be added using typical conversion efficiencies between primary to final energy for different energy carriers	Territorial Example: Regional energy statistics, Baynes and Bai (2009)
Regional energy metabolism	Physical	The city as socio-metabolic system	Combination of final, secondary, and primary	Not included, but can be added using typical conversion efficiencies	Territorial Example: Schulz (2007)
Regional economic activity	Economic (physical extensions)	Final demand (economic)	Total Primary Energy Supply	Includes upstream and embodied energy of goods and services, no sectorial differentiation	Territorial Example: Dhakal (2009)
Energy Input–Output	Economic (physical extensions)	Final demand (economic)	Total Primary Energy Supply in national analysis, primary energy in global analysis	Includes upstream and embodied energy of goods and services, sectorial differentiation	Consumption Example: Wiedenhofer et al. (2011), Weisz et al. (2012)

consumer of final goods and services in economic analysis which treats firms as intermediate consumers and reserves the term final consumer (consumption) to household and governmental consumption, capital formation and exports. These sectorial accounts of urban final energy use also allow comparisons with national level data or data of other cities and can serve as a useful guide, e.g., for energy-efficiency 'benchmarking' that can guide policy.

The direct final energy account can further be extended by estimating the *upstream energy* requirements needed to provide the final energy, using e.g. lifecycle analysis. The upstream energy is the primary and/or secondary energy use *linked to the final energy* utilized by the end users within a city. Depending on the estimation method, the upstream energy may or may not include the energy required to extract and transport the primary energy itself. For clarity, it is crucial to specify which type of upstream energy is considered: secondary or primary, including energy for extraction activities themselves. To avoid confusion with other terms used for upstream energy, this section reserves the term 'upstream' for the primary or secondary energy needed to generate final energy, used in the city and the term 'embodied' for the primary or secondary energy needed to produce all other goods and services consumed in the city, regardless of where this energy was mobilized, within or outside the city boundaries.

Final and upstream energy uses are not the total primary energy required by urban activities, since they do not include the energy needed to produce goods and services *consumed* in the city, but imported from outside. Conversely, the final and upstream energy uses of a city also include energy uses to manufacture goods and provision of services which may be consumed in the city or else be *exported* from the city (and consumed in other cities or in rural areas). Care therefore needs to be taken to avoid double counting of energy flows, for example by adding imported 'embodied' energy flows to the final and upstream energy uses of a city, but ignoring the (final and upstream) energy embodied in goods and services produced in a city, but exported for consumption elsewhere. This double counting is averted in input–output methods, but can be a problem in approaches that rely on life-cycle analysis (Ramaswami et al. 2008; Hillman and Ramaswami 2010).

The difference between the final and upstream energy use can be considerable. Take the example of electricity: typically, for 1 GJ of electricity consumed in a city, up to 3 GJ of primary energy in the form of coal must be burned in a conventional steam power plant. (This ratio is substantially lower for combined-cycle power plants fired by natural gas, which illustrates the need to consider the *actual* urban energy system characteristics rather than aggregate 'upstream' adjustment factors.) Heating fuels, such as gas and fuel oil, also have upstream energy use through their extraction and transport. According to Kennedy et al. (2009), who applied this method to the GHG emissions of ten global cities, the lifecycle GHG emissions associated with urban fuel use are between 9 percent and 25 percent higher than their local emissions. This approach is also that followed by the Harmonized Emissions Analysis

Tool of the International Council for Local Environmental Initiatives (ICLEI 2009), although since the software is proprietary (and not transparent), geographically limited, and often applied only to municipal energy use, it is not clear how reliable or significant the results are.

The 'regional energy metabolism' method also uses physical data, but defines the user as a socio-metabolic system (in this case, the urban-metabolic system) and measures all energy flows that cross the boundary of the urban system, regardless of their use and conversion stage. The method applies the more general concept of socio-economic metabolism, which includes energy, materials, water, waste flows, emissions, infrastructure stocks, and other elements of the built environment to cities. The system boundary is defined twofold: as boundary towards other socio-economic systems (the political boundary) and towards the natural environment within the political territory (Haberl 2001). Consequently two types of flows cross the boundary of the urban system: imports from and exports to other socio-economic systems, and extraction of raw materials from and release of emissions and wastes to the natural environment within the city's territory. Because primary extraction activities (such as farming or mining) predominantly take place outside the political city boundaries, the typical urban metabolism is dominated by import and export flows, whereas the socio-economic metabolism of larger regions (such as nation-states) typically is dominated by the domestic extraction of raw materials and the release of wastes and emissions (Matthews et al. 2000; Weisz et al. 2006; UNEP 2011). A complete urban energy metabolism would include the calorific value of materials extracted form the city's territory (by e.g. city farming) and of all physical goods entering and leaving the city including biomass for human (and animal) nutrition, energy flows which are not included in other approaches.

In contrast to the final energy method, the primary energy consumed by the energy sector *within the city* is included in the regional energy metabolism. Thus, the location of a power plant inside or outside a city's boundaries has a large influence on the measurement of its energy consumption in this method. As a result, the regional energy metabolism of a city is always larger than its final energy. The mixture of primary and final energy, however, makes it difficult to compare this method's results with national level data and those of other cities.

So far, the urban metabolism approach has been applied to material flows rather than energy flows. Examples include Cape Town (Gasson 2007), Geneva (Faist et al. 2003), Hong Kong (Newcombe et al. 1978), Paris (Barles 2009), and Singapore (Schulz 2006, 2010a). Table 4.2 summarizes some findings from urban metabolism studies. These studies highlight that the urban energy metabolism has a significant material dimension: in fact material energy carriers (fossil fuels and biomass) represent a large share (a quarter to a half) of domestic material consumption, an aggregated indicator that measures all materials inputs (except water), minus the amount of exported goods (all measured in tons). Water is kept separate because its quantity exceeds that of all other materials by at least a factor of 10.

Table 4.2 Energy and material flows of selected cities showing the importance of energy flows in the total metabolism of cities

City	Cape Town	Geneva	Hong Kong	Paris Petite Couronne	Singapore
Reference	Gasson (2007)	Faist et al. (2003)	Newcombe et al. (1978)	Barles (2009)	Schulz (2005, 2010a)
Population (millions)	3	0.4	3.9	6.3	4.1
Year	2000	2000	1971	2003	2000–2003
Energy (GJ/cap/year)					
Primary	40		72		258
Regional (city)	22	92	43		103
Domestic material consumption (tons/cap/year)					
Total		7.7		4.6	29.7
Fossil fuels		1.8		2.1	5.2
Biomass		1.0			0.3
Construction minerals		4.8			22.5
Industrial minerals and ores		0.1			1.7
Water	110	151	100		112

* Domestic material consumption = domestic extraction + imports – exports, which in the urban case is dominated by imports.

The potential of the urban metabolism approach has not been fully explored yet. With more standardized urban specific data and more complete and comparable empirical examples urban metabolism studies could contribute to several aspects of the urban energy system: the role of cities as exporters of goods and its associated energy requirements, understanding the mutual relationships between stocks and flows, and between energy and material uses (the extraction, transport, transformation, and maintenance of the material components of the urban-metabolic system requires energy, the provision and distribution of energy services requires materials as energy carriers and for energy infrastructure), the inclusion of biomass as an element of the urban energy system would allow more meaningful comparisons between industrial and pre-industrial cities, and its related land use consequences, the inclusion of upstream and embodied energy would greatly enhance comparability with the national level and between cities.

The 'regional economic activity' approach uses the urban final demand aggregate (defined in *economic* (monetary) terms i.e., as gross regional product (GRP)) and a national or regional energy-to-GDP relationship to estimate the city-specific energy use. This method was used by Dhakal (2009) for major Chinese urban areas (see Chapter 9) and is the only possible approach when no city-specific energy data are available. However, the limitation of this method is that it ignores all city-specific drivers of urban energy use except income. The national or regional energy-to-GDP ratio usually uses total primary energy supply as the energy variable. This method is thus fully comparable with statistics on the national level.

'Energy I–O' analysis is based on economic input–output tables, which measure (usually in monetary terms) all sales and purchases of goods and services among the producing sectors and to final demand. These tables can be extended to account for physical energy flows or emissions. Using specific linear algebra equations, the 'embodied' energy (i.e., the energy used throughout the whole production chain to produce the final goods and services) can be calculated from these tables. This approach allows the energy requirements of the producing sectors to be allocated to the final consumption sectors (i.e. household consumption, governmental consumption, capital investments, and exports). Hence, this approach is often referred to as a 'consumption' approach, as opposed to the 'production' approach that focuses on apparent energy use. As the underlying input–output tables are usually national tables, the national final demand vectors are disaggregated using consumer-expenditure surveys for urban areas. In practice most applications of the input–output method to urban energy use only consider urban household consumption, which is in most cases also the largest urban final consumption category.

The major advantages of the input–output method are the high sectorial resolution of the production and final consumption systems (ranging approximately from 20 to 500 different sectors, depending on the quality of the national database) and a mathematical allocation algorithm that avoids doubling counting as well as gaps in the allocation of energy data.

This energy input–output method has been used to assess the energy consumption of Indian urban and rural populations (Pachauri et al. 2004; Pachauri 2007), Brazilian urban households (Cohen et al. 2005), Sydney's energy use (Lenzen et al. 2004), as well as for postal district resolution maps of Australia (Dey et al. 2007).

Illustrative results on direct versus embodied energy use in Asian megacities based on I–O analyses are given in Figure 4.1. First, it is important to recognize how embodied energy flows compare with final energy flows, particularly for high-income cities such as Tokyo. High-income cities, whose economies tend to be service based rather than industry based, tend to import large quantities of embodied energy, compared to the energy required directly for urban operations. Physical urban growth will tend to lead to increases in the energy required for urban operations (regional energy), whereas growth in economic exchanges, especially in expenditures, will lead to growth in embodied energy. In the Asian megacities shown in Figure 4.1, the urban growth effect on direct energy use is larger than changes in trade flows into the city over the limited time period considered in the study.

The differences between consumption and production energy flows are shown schematically in Figure 4.2. The input–output-based 'consumption' approach can differentiate between cities with different incomes and final consumption patterns, and between the sectorial production technologies of its national – or in the case of multiregional input–output models – its global – hinterland. It cannot, however, capture urban specificities in industrial and service energy use, and specifities of the urban energy supply systems (e.g., degree of cogeneration) are only captured indirectly via comparably less expenditure for transportation and heating in cities with efficient energy supply systems. With the use of urban consumption surveys, the city specific consumption patterns are well represented in these studies. However, the assumption of homogeneous sectorial prices, inherent in any input–output analysis based on monetary I–O tables, can introduce a specific bias to the allocation of energy flows to urban consumption in cases where consumer prices in urban areas are significantly higher than in the whole country on average. This problem could in principle be overcome by using input–output tables in physical units (Proops 1977; Weisz and Duchin 2006), however these tables are difficult and time-consuming to compile and therefore hardly available. A less time-

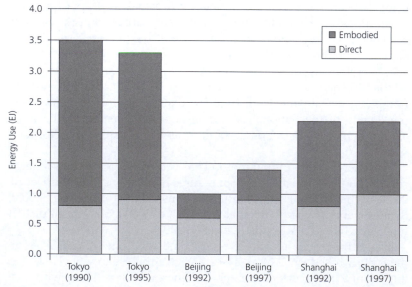

Figure 4.1 Estimates of direct (final) versus embodied (via imports of goods and services) energy use (in EU) of Asian megacities
Source: Dhakal (2004)

consuming solution for energy input–output analysis is the use of mixed unit tables, where the output of the energy sectors is measured in Joules and the output of all other sectors is measured in monetary units. This approach was for example applied by Pachauri and Spreng in their 2002 study of direct and indirect energy consumption of households in India. Nonetheless, issues of price differences and of assumed product homogeneity in the input–output method have not been addressed systematically in the context of urban energy analysis.

National I–O models do not provide information on embodied energy flows in commodities and goods traded internationally, which requires the use of multiregional I–O models. Studies performed for Norway (Peters et al. 2004), the United Kingdom (Wiedmann et al. 2007), and the United States (Weber and Matthews 2008) show that a significant fraction of energy use can be attributed to the embodied energy of imports, especially in industrial economies (Peters and Hertwich 2008). The application of the I–O method to developing countries is, however, not straightforward, since the large informal sector is absent from the official I–O tables, which themselves exist in up-to-date versions only for a few developing countries. It has also been shown that uncertainties are very large for the GTAP (Global Trade Analysis Project) database used in multiregional I–O models (Weber and Matthews 2008).

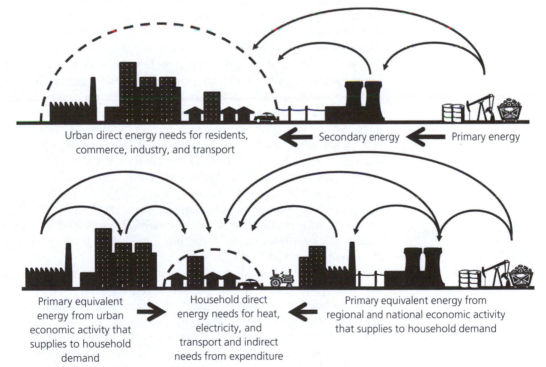

Figure 4.2 Comparison of scope for the regional production approach (top) and household consumption approach (bottom). Above-ground arrows show path of direct or indirect energy use, lower-level arrows show the general progress of primary energy to the end-users
Source: Figure courtesy of Baynes et al. (2011)

From the I–O studies, the largest consumption categories in urban household energy use are housing, transportation, and nutrition. In the housing category the bulk of energy is used for heating. Among the final energy types used to serve the energy demands in these three consumption categories, electricity and biomass (food) have the largest upstream and embodied energy 'content'. Thermal electricity generation involves heat losses up to two-thirds during transformation, plus transmission and distribution losses. Food has both nutritional and embodied energy components: a diet of 3,000 kcal/day corresponds to 4.5 GJ/year in nutritional energy per person. The commercial primary energy required to produce food ranges from 2.5 to 4.0 GJ/capita for Indian urban households and from 6 to 30 GJ/household for Brazilian urban households (where the ranges correspond to low and high income brackets) to around 40 GJ/capita for European households (Vringer and Blok 1995; Pachauri 2004; Cohen et al. 2005). The majority of the energy embodied in food production is consumed outside city boundaries.

4.3 Comparison of energy-accounting frameworks

To compare and contrast the results from different accounting approaches we initiated a collaboration among various research teams to provide a quantitative illustration, applying two different methodologies ('final energy' and 'energy I–O') for two different cities: London and Melbourne. Recent (partial) results for Beijing are also included for comparison.

For the Melbourne study, Manfred Lenzen and his group at the University of Sydney used environmentally extended I–O methods coupled with household expenditure surveys to map direct and embodied GHG emissions of household consumption (Dey et al. 2007). This method was adapted to provide results in terms of primary energy use for the city of Melbourne (Australia). Baynes and Bai (2009) scaled state data down to the urban level, focusing on the (direct) final energy use of Melbourne city. The Melbourne comparison and methodology are described in detail in Baynes et al. (2011). The London study compares final energy use from official statistics (UKDECC 2010) and results from a multiregional, environmentally extended I–O analysis with explicit representation of the household consumption vectors for the Greater London Authority (GLA).

The yet unpublished study is based on a method of Minx et al. (2009) and was carried out by TU Berlin, PIK and SEI (Weisz et al., 2012). The Beijing study (Arvesen et al., 2010) only considered household energy use (and hence misses the large industrial and service sector energy use) and combined both the final energy method (with additional approximate fuel-specific estimates of upstream conversion energy needs) with an I–O approach.

The results are summarized in Table 4.3. To enable comparison, the position along the energy chain has to be the same, so the primary energy equivalent of final energy is estimated where detailed statistics are unavailable (using standard conversion efficiency factors from Kennedy et al. 2010). To correct for different sizes of the cities, all values are expressed in GJ/capita.

The two methods cover different types of energy flows, some of which can be compared, and others cannot. Energy I–O ('consumption accounting') focuses on the energy use of households within the city boundary directly and embodied in the purchase of goods and services. A direct comparison with the territorial final energy ('production accounting') method is only possible, therefore, for the energy directly purchased by households: for residential housing (heating and electricity) and (in the case of Melbourne and Beijing) private transportation. The final energy method also measures urban nonresidential energy (for industry and commercial activities, as well as non-private transportation), which the energy I–O method does not cover. Conversely, the energy embodied in the household purchase of goods and services is not covered by final energy method.

The most interesting result lies in the (first ever) quantification of the differences between the two different accounting methods. As expected, the consumption-based accounting method yields much higher total energy use for London (+30 percent)[5] and Beijing (+100 percent) than the production accounting method. Conversely, Melbourne's territorial energy use is significantly higher than the energy consumed directly and indirectly by its households. This is not because the Melbourne households consume less energy: in fact, in total, they consume almost one-third more than the London households, mainly through private

Table 4.3 Primary energy use for two different energy-accounting approaches for three cities for which (partial) data are available: Melbourne, London, and Beijing. All values are expressed in GJ/capita (permanent) resident population. Dashes (–) indicate categories of energy use that cannot be compared directly between the two different accounting methods

Primary energy GJ/capita for:	Melbourne 2001		Greater London 2004		Beijing 2007 Household energy only	
	Pop.: 3.2 million		Pop.: 7.6 million		Pop.: 12.1 million	
	Prod. acc.	Cons. acc.	Prod. acc.	Cons. acc.	Prod. acc.	Cons. acc.
Residential heating	22	12	28		9	
Residential electricity	28	30	17		9	
Residential housing (heating + electricity)	50	42	45	35	18	11
Private cars	33–41	27	10		7	7
Nonresidential use	197	–	56	–	n.a.	–
Household consumption of goods and services	–	116	–	108	–	34
Total	279–288	184	111	143	25	52

Pop = population; Prod. acc. = production accounting; Cons. acc. = consumption accounting; n.a. = not available. Numbers may not add up exactly due to independent rounding.

transportation (cars). Instead, it is because Melbourne's non-residential energy use is almost quadruple that of London's, on a per capita basis. Melbourne is still a major industrial production center, and this industrial activity results in high industrial energy use in the production account (three times as much as household energy use in the production account). London as defined by the GLA, in contrast, has very little industrial activity, with services dominating its economic activities, and so household consumption is larger than the territorial production-account energy use. The importance of industrial energy use is also illustrated in the case of Beijing. Total secondary energy use (all sectors) in Beijing in 2007 was some 145 GJ/capita (Beijing Government 2010), i.e., three times larger than the total direct plus embodied estimated household energy use reported in Table 4.3. Taking upstream energy conversion losses into account, the primary energy use (the energy level directly comparable to the other cities in Table 4.3) of Beijing is approximately 200 GJ/capita, i.e., in the same ballpark as Melbourne or London (production accounting). The major difference is that average per capita income in Beijing is with 10,000 *purchasing power parity* dollars (PPP$) per capita, approximately a factor of four lower than that of Melbourne and a factor of six lower than London's, suggesting the twin importance of economic structure and efficiency of energy end use as determinants of urban energy-use levels, with the latter offering a substantial potential for improvement.

The above results confirm that production accounting of energy reflects the economic structure of urban areas, and their role in the international division of labor, whereas consumption accounting energy reflects a mixture of local conditions (climate and transit infrastructure) and expenditure levels (income and lifestyle effects). This exercise also demonstrates the power of applying and comparing different methods at the urban level. By showcasing the differences in production and consumption energy, the potential role of local policy measures (e.g., transport) versus broader consumption measures (consumption reduction, or low energy/emissions supply chain management) can be made explicit.

The comparison of the two methods indicates two policy avenues. Local policy priorities should focus on housing, transit, and industrial energy savings. But these must imperatively be complemented by shifts and reductions in the energy embodied in household purchases of goods and services to avoid that savings at the regional level are offset by increased energy demand from the consumption of goods and services, which occurs somewhere else in the world. Such a policy agenda clearly goes beyond the local level and needs to be addressed on multiple scales (Ramaswami et al. 2011; Baynes et al. 2011).

4.4 Spatially explicit urban energy accounts

Spatially explicit energy-use studies may be the key to understanding the influence of urban form, and periurban and urban specificities. The spatial disaggregation can again be based on a consumption account or else on a production (territorial) account as described above. For Sydney, Lenzen et al. (2004) disaggregated consumption-based energy use in fourteen areas, and followed this up with a GHG emissions atlas of Australia at the postal district level (Dey et al. 2007). For Toronto, VandeWeghe and Kennedy (2007) derived spatially explicit production-based energy use data from transportation and energy-expenditure surveys (see Figure 4.3). Andrews (2008) analyzed production-based energy use of several districts in New Jersey, ranging from rural to urban. A comprehensive spatial GHG account, including discussions of uncertainties, was recently completed for the city of Lviv, Ukraine. An innovative study used the US Vulcan emissions atlas[6] to compare transportation and building emissions in urban, periurban, and rural counties of the United States (Parshall et al. 2010).

The Sydney and Toronto studies found higher per capita energy and emissions in the outer, low-density suburbs. The New Jersey study also found a 'threshold' in per capita urban transportation energy compared to more rural counties. In Sydney and Toronto, building fuel use was higher in the city center. The causes for higher building energy in the center of the city could be the age and quality of the housing stock, the presence of an energy-intensive central business district, and higher incomes in those areas. In the Sydney study, where economic information is available, building energy use is highly correlated with income, but less correlated with population density. In Sydney, central districts tended to have higher incomes

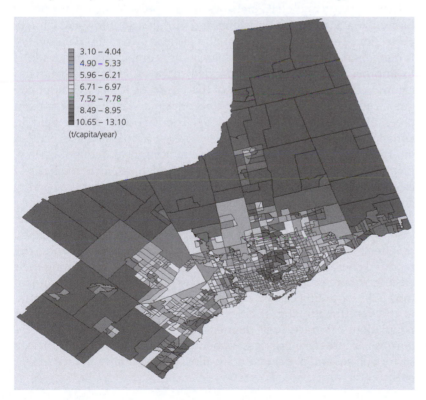

3.10 – 4.04	
4.90 – 5.33	
5.96 – 6.21	
6.71 – 6.97	
7.52 – 7.78	
8.49 – 8.95	
10.65 – 13.10	

(t/capita/year)

Figure 4.3 Total GHG emissions from Toronto (tons CO_2-equivalent/capita/year). High-resolution images as well as maps for various energy-demand subcategories (residential, transport, etc.) are available from VandeWeghe and Kennedy (2007). (See color plate 5)

than the outer suburbs, a trend absent in the New Jersey study. Since these studies are of industrialized countries and automobile-based urban areas of North America and Australia, their results may not apply to urban areas more generally.

4.5 Recommendations for urban energy assessments

As urban energy statistics have a vital role in allowing agents to benchmark, it is essential to be sure that the methods are comparable and that 'gaming' is not taking place. Each of the methods described above can produce results that allow benchmarking and comparisons, as long as the method and data sources are described clearly, and consistent data, sectorial definitions, and system boundaries are applied and clearly spelled out (see Box 4.1). The method should also be as transparent and as open as possible, to guarantee reproducibility and fact checking. Moreover, examples of energy assessments that only account for some sectors are not measurements of urban energy. The sectorial distribution of energy use (residential, commercial, industrial, administrative), the contribution of different energy services (mobility, indoor temperature, nutrition, other goods and services) as well as the final energy mix (electricity, petroleum, gas, biomass, etc.) are essential complementary elements for informed analysis and decision making and should be an integral part of urban energy reporting.

Urban energy and GHG statistics should provide a basis for policy formulation, investment decisions, and further action towards climate protection. Therefore, it is essential that their origin and data quality is made transparent and methodologies are comparable. Suggestions to improve terminology are provided in this chapter. It is far too common to read 'the energy use of this city is XX Joules', without any qualification what type of energy (final or primary, including upstream or embodied energy) is referred to. There is also a rich field in enhancing the usefulness of urban energy accounts by expanded information on energy quality (e.g., separating heat demand by low, medium, and high temperature regimes), which can form the basis of extended energy efficiency studies, for example in the form of exergy analysis (see Box 9.2 below).

City energy assessments should also include clear definitions of the system boundary used. Currently, many urban energy assessments, in effect, arbitrarily choose the system boundary to reduce the reported energy use or GHG emissions, for instance by claiming their electricity comes from different sources than the average regional mix, or by excluding certain energy uses that are, nonetheless, central to the very functioning of cities, such as airports or a large tourist population. Arbitrary, or ill-defined, system boundaries defy the very purpose of urban energy assessments: to guide public and private sector policies and decisions and to allow comparability and credibility of the entire process.

Box 4.1 Urban energy data: measurement and quality issues

For urban energy data and assessments two major issues need to be spelled out in a clear and transparent way: (1) system boundaries, and (2) data availability and quality issues.

(1) Within the discussion of system boundaries two issues need to be considered:

  (a) What is the spatial or functional definition of the urban system under consideration? Does the city definition refer to the core city alone, or does the assessment include the larger metropolitan area? Does the system definition include recognition of bunker fuels[7] (transport fuels used outside of the spatial system boundary, e.g. in national and international territory) or not? Does it consider also the embodied energy associated with the use of material resources and goods other than energy carriers, or not?

  (b) What is the energy system considered? Is primary or final energy reported, and to what extent is a lifecycle perspective for the fuel provision followed (e.g., upstream energy conversion losses and associated emissions, or the costs of exploration, drilling, transporting, and refining fuels before import into the urban system are included or omitted in the analysis)? How are airports, harbors, commuting beyond the urban boundary, fuel bunkers, reserves, and so on addressed? What about decentralized generation technologies, such as air pumps, geothermal and combined heat and power (CHP)? These may not be important in all cases, but they should be at least considered.

(2) Quality and availability of energy data: are actual statistics used or extrapolated/downscaled data? Does the assessment include noncommercial energy?[8] Which spatial and temporal resolution was considered to calculate the fuel mix for electricity provision? Are differences in technology, efficiency, etc., of power plants and other energy conversion processes recognized?

In an ideal world, urban energy reporting should adopt as wide systems boundaries and complementary accounting frameworks as is reasonably possible and available data allow.

When narrower system boundaries are adopted, a simplified sensitivity analysis of the effects of inclusion of omitted system components can help to put reported numbers into a proper perspective (i.e., complementing final energy accounts with estimates of corresponding primary energy needs, or production-based accounts by estimates of consumption-based accounts based on national I–O tables).

Incomplete reporting (e.g., of only municipal energy use) should be avoided as only a comprehensive sectorial perspective of all urban energy uses can reveal the full potential for policy intervention and assure comparability across different urban energy accounts.

Finally, data disclosure and documentation of assumptions and methods are a 'must'. Particularly, the area of urban GHG inventories is replete with examples of glossy policy briefs that do not allow the reproducibility of the numbers presented (not to mention unreported uncertainty ranges). Transparency and data disclosure are not only key from the perspective of scientific integrity, quality, and reproducibility, but they are also the key for well-informed policy choices. A comparable effort to the standardization of energy and GHG accounts at the national scale along the OECD/IPCC model is long overdue for the urban scale as well.

Notes

1   The distinction between a 'production' and a 'consumption' approach originates from discussions about appropriate system boundaries for national CO_2 inventories. The question was if CO_2 emissions should be allocated to the country where they were produced by industries and households (producer/production or territorial principle), or if they should be allocated to the country where the final consumption takes place (consumer/consumption principle) (Peters et al., 2004, Peters and Hertwich 2008). Here we apply the same terminology to urban energy use instead of national CO_2 emissions. Although the term production for direct energy use is somehow misleading, the original terms (well established in the national scale literature) are retained here to avoid introducing further terminological complexity.

2   The terms 'direct' and 'indirect' energy are avoided in this chapter, since they have a different meaning in each of the approaches considered. Instead, the terms 'final', 'primary', 'upstream', and 'embodied' energy are used (see the Glossary in Chapter 1 for a definition of these terms and concepts).

3   The term 'urban metabolism', first introduced by Wolman (1965) refers to the flows of energy, materials, and water that enter an urban area from outside, their internal transformation into stocks, goods, services, and wastes, and their final release into the environment of the urban system as wastes, emissions, and exported goods. These flows are measured in physical units (tons or GJ) to allow calculating mass and energy input–output balances of urban systems.

4   The ecological footprint is an aggregate measure generated by converting the amounts of resources used and emissions generated from human consumption at different scales (global, national, urban, or individual) into the required global hectares of bioproductive land, assuming a globally uniform average land productivity (therefore the term 'global hectares'). As a consequence of using bioproductive land as the normalizing standard, the ecological footprint is heavily biased towards CO_2 emissions (and their required land-based CO_2 sinks) and biomass use quantities and land productivities.

5   CO_2 accounts for London (Hersey et al. 2009) suggest that the differences between the two accounting methods could be 100 per cent (some 45 versus 90 million tons CO_2 for the production versus the consumption accounting, respectively), a difference that appears very large in view of the results from the energy comparison reported here. I–O techniques are used here (Table 4.3) whereas the London CO_2 study used a life-cycle assessment approach to estimate consumption-based CO_2 emissions (but the underlying method and data have not been published), so there might be an additional methodological explanation for the large differences in estimates of the energy and CO_2 consumption-accounting 'footprint' of London.

6   Vulcan is a gridded data inventory of fossil fuel based sources of carbon emissions, covering the entire United States. The spatial resolution is sufficient to distinguish between urban and non-urban emissions. The Vulcan database is maintained by Purdue University (available online at: www.purdue.edu/eas/carbon/vulcan/research.html).

7   Bunker fuels, i.e. energy used for (international) air transport and shipping, can be a substantial fraction of urban energy use. In 'world cities', such as London and New York, air and maritime bunker fuels can account for about one-third of the direct final energy use, suggesting the importance of their inclusion in urban energy accounts.

8   For many cities in developing countries, non-commercial energy forms can account for a substantial fraction of urban energy use (one-third to one-half). Its reporting is therefore not only key for a comprehensive urban energy accounting, but also yields important information on the potential of fuel substitution with rising urban incomes and hence future energy infrastructure needs.

5
Urban energy use

Arnulf Grubler and **Niels B. Schulz**

5.1 Current urban energy use (global and regional)

How large is the urban fraction of global energy use? This seemingly simple question is hard to answer as, contrary to the data for countries, no comprehensive statistical compilation of urban energy use data exists for whatever system boundary one might consider in the analysis. With 50 percent of the world population being urban, a range of (largely ballpark) estimates put the urban energy share between two-thirds to three-quarters of global energy use, but such global estimates have, until recently, not been supported by more detailed assessments. This section reviews the two detailed assessments of urban energy use available to date: the estimate of the IEA published in its 2008 World Energy Outlook (IEA 2008) as well as an estimate developed by a team of researchers at the International Institute for Applied Systems Analysis (IIASA) for the Global Energy Assessment (GEA).

In the absence of detailed, comprehensive urban energy-use statistics, two analytical approaches were pursued to derive global (and regional) urban energy use estimates. One technique, which might be labeled 'upscaling', uses a limited number of national or regional estimates of urban energy use and then extrapolates these results to the global level. This is the approach followed by the IEA (2008) study that estimated (direct) urban energy use (based on direct final energy use and reported for an inferred primary energy total at an urban scale). The second approach adopts 'downscaling' techniques in which national level statistics are 'downscaled' to the grid-cell level, and then combined with geographic information system (GIS)-based data sets on urban extents to derive spatially explicit estimates of urban energy use. These are then aggregated to the national, regional, and global levels. This approach underlies the IIASA study that estimated urban energy use at the level of (direct) final energy.

In the 2008 *World Energy Outlook of the International Energy Agency* (IEA 2008) a separate chapter is devoted to urban energy use and contains estimates of 2006 base-year urban primary energy use data and a reference scenario projection to 2030. Detailed urban energy use assessments were first commissioned for a limited number of countries and regions (China, the United States, the European Union, and Australasia (i.e., Australia and New Zealand)). In these regions urban energy use is estimated to range from 69 percent (European Union) to 80 percent (United States) of the primary energy use of these regions, which reflects their high degree of urbanization. For China the estimate is 75 percent, despite a comparatively lower urbanization rate (41 percent compared to 81 percent in the United States), but explained by the substantially higher urban energy use in Chinese cities compared to the national average because of higher urban incomes and urban industrial activities. The results of the 'upscaling' of these four regional sets of data to the global level are not reported separately by region by IEA, so only global totals can be discussed here.

The IEA (2008) estimates urban primary energy use at the global level to amount to some 330 EJ for the year 2006, or 67 percent of world primary energy use. Using an average global primary-to-final energy conversion efficiency of about 69 percent, the estimate translates to 230 EJ urban final energy use worldwide, which is in good agreement with the IIASA study results reported below. Estimates are also provided by major primary energy source and for electricity, assuming that the primary energy mix of cities is the same as at the national or regional average. This assumption is problematic, especially for countries in low-income, low-urbanization regions, such as Asia and Africa, where available data suggest that urban energy use structures are, in fact, very different from rural and national averages. Urban energy use is invariably characterized by much higher shares of grid-dependent energy carriers (electricity and gas) and by much lower reliance on traditional biomass fuels. This simplifying assumption in the IEA (2008) study also diminishes the plausibility of the study's scenario projections by primary energy carrier to 2030. Total urban primary energy use is projected to grow by some 56 percent from 2006 to 2030. In the IEA reference scenario almost 90 percent of global energy growth to 2030 is projected to result from increased urban energy use.

The IIASA study follows a different approach. Drawing on methods and data sets (see Grubler et al. 2007) developed for spatially explicit GHG emission scenarios, the IIASA study used spatially explicit GIS data sets of urban extents, constrained to be consistent with the latest UN World Urbanization Prospects (UN DESA 2010) statistics for the year 2005 as initial input. In a subsequent step, national-level final energy use data by fuel (traditional biomass and electricity) as well as by end-use activity (primary, light and heavy manufacturing industries, households, and transportation) were downscaled to the grid-cell level in proportion to available spatially explicit activity variables[1] (population, GDP, light luminosity, etc.) under a range of plausible algorithms (hence the study provides central as well as minima/maxima estimates to illustrate uncertainty). The scenarios of individual final energy use categories were then aggregated per individual grid cell and overlaid with the urban extent map to derive the total estimated (direct) final energy use (including noncommercial traditional biomass fuels) in urban areas. Table 5.1 summarizes the results for the eleven GEA regions and five GEA world regions, as well as for the global total.

Globally, urban final energy use in the IIASA study is estimated to range from 180 to 250 EJ with a central estimate of 240 EJ, or between 56 percent and 78 percent (central estimate: 76 percent) of total final energy. So, in terms of final energy use (as opposed to primary energy use reported in the IEA study), cities use 240 EJ, or some three-quarters, of final energy worldwide. The absolute amounts are in good agreement with the IEA (2008) study discussed above, at least globally.[2] Readers should not be confused by the somewhat higher urban percentage (76 percent) of urban *final energy* use of the IIASA study when compared to the 67 percent estimate of the IEA for *primary energy* use. As discussed above, the assumed identity in urban fuel and energy mix with national and/or regional averages in the IEA study underestimates the level of high-quality, processed-energy forms in urban areas that entail correspondingly higher upstream energy-conversion losses. If this simplifying assumption in the IEA calculations would be relaxed, the corresponding urban primary energy estimate would become higher and much closer to the three-quarter benchmark of the IIASA study (i.e. urban primary and final energy use fractions in the two estimation approaches would converge).

These observations are corroborated by commercial final energy use estimates in urban areas, i.e., excluding traditional biomass use. For industrialized countries, estimates of urban commercial fuel use are identical to the totals reported in Table 5.1. Major differences exist, however, for some developing regions. For sub-Saharan Africa, estimates suggest that 4 EJ, or some 80 percent of all commercial energy use, can be classified as urban (compared to 8 EJ and 54 percent for total final energy including noncommercial energy carriers such as traditional biomass; see Table 5.1). Differences for South Asia are also noticeable: 8 EJ, or 71 percent of final commercial energy, are classified as urban, compared to 10 EJ and 51 percent for total final energy (including noncommercial energy). Differences for the other developing GEA regions are comparatively minor, as little noncommercial energy continues to be used in cities. The higher urban share in commercial energy results both from higher urban incomes and better urban energy access and infrastructure endowments, particularly the much higher degrees of electrification in urban areas.

Nonetheless, despite some uncertainties[3] (see Table 5.1), both the IEA and the IIASA estimates confirm a highly policy-relevant finding: While some 50 percent of the world's population is urban, *urban energy already dominates global energy use*, which means that the energy sustainability challenges need to be solved predominantly for urban systems.

5.2 GEA city energy data and analysis

5.2.1 The GEA city energy database

An effort to compile a database with literature values of energy use at the urban scale was conducted as part of the GEA assessment to

Table 5.1 Estimates of urban (direct) final energy use (including traditional biomass) for the GEA regions and the world in 2005 (in EJ and % of total final energy). See text for a discussion of urban commercial energy use and its corresponding (somewhat higher) urban share

GEA Regions		central estimate		min		max	
		EJ	as %	EJ	%	EJ	%
NAM	North America	63	86	51	69	64	87
PAO	Pacific OECD	14	78	11	59	16	92
WEU	Western Europe	40	81	31	64	41	83
EEU	Eastern Europe	6	72	4	51	6	72
FSU	Former USSR	20	78	14	54	20	78
AFR	Sub-Saharan Africa	8	54	5	35	10	71
LAM	Latin America	17	85	16	77	18	89
MEA	North Africa & Middle East	15	84	10	58	15	86
CPA	China & CP Asia	32	65	19	40	31	65
PAS	Pacific Asia	15	75	10	51	16	77
SAS	South Asia	10	51	5	29	10	51
OECD90	NAM+POA+WEU	117	83	92	66	121	86
REFs	EEU+FSU	25	76	18	54	25	76
MAF	AFR+MEA	22	71	15	47	25	79
LAC	LAM	17	85	16	77	18	89
ASIA	CPA+PAS+SAS	57	64	35	40	57	64
WORLD		239	76	176	56	246	78

improve understanding of the variation in energy demand of urban areas (for an example of such analyses, see Steemers 2003). The study, therefore, chose a cross-sectorial approach to compare as large a number of urban areas from as wide range of regional settings, geographies, sizes, and functions as possible, with minimal definitional constraints with respect to urban system boundaries so as to maximize data availability. In terms of energy use, data are reported at the level of *total (direct) final energy use*, as this level of analysis creates the least ambiguity in terms of energy accounting and is also the indicator most widely available and comparable among case studies. (Accounting for primary energy equivalents or GHG emissions requires assumptions on boundary definitions, conversion factors, and efficiencies, etc., which introduce additional uncertainties in the comparisons.) Given the extreme paucity of consumption-based estimates of urban energy use (e.g., via I–O techniques), the decision to focus the database on a production approach was also straightforward.

5.2.2 Data coverage

Three categories of urban statistical data were brought together in the GEA city energy database from a variety of sources: population statistics (UN 2008), energy statistics (e.g. Dhakal 2009; Kennedy et al. 2009, 2010), and economic statistics on gross regional economic output (or GRP, which is the urban-scale equivalent of national GDP) converted into a common 2005 denominator in purchasing power parity (PPP expressed in International $ – Int. $2005) terms, including Eurostat 2008 and PriceWaterhouseCoopers 2007. While population statistics are routinely collected at various levels of spatial resolution, this is rarely the case for both economic and energy consumption data. Coherent data sets were, nonetheless, found for 225 urban units, of which 160 were from UNFCC Annex 1 (i.e., industrialized) countries and 65 were urban areas located in non-Annex-I (i.e., developing) countries. Details on data-source limitations, as well as further statistical analysis, are reported in Schulz 2010b. In the following analysis, urban areas are grouped into Annex-I[4] ('industrialized') and non-Annex-I countries ('developing'). The Annex-I countries are further subdivided in OECD 90 countries (the traditional high income countries) and those who more recently joined (REF, mainly the economics in transition from planned to market economies in the former USSR and in Central and Eastern Europe). Non-Annex-I countries are subdivided into Non-OECD Asia on the one hand, and the aggregate of Latin America, Middle East and Africa on the other. The total population covered in this dataset is 480 million (17 percent of the global urban population in 2000), the total final energy consumption reported amounts to 42 EJ, about 15 percent of the global final energy use.

5.2.3 Analysis

5.2.3.1 Comparisons of urban-scale and national-scale data

This section compares data on (urban) energy use per capita versus per capita income (GRP/GDP), and energy intensity of GRP/GDP at the urban scale with their respective national-scale metrics.

Table 5.2[5] presents the overall results and a regional breakdown by status regarding Annex-I versus non-Annex-I designation under the UN Framework Convention on Climate Change and by geographic regions. Demographic data at the national scale are derived from UN DESA 2008 and 2010, national energy statistics from IEA energy balances (IEA 2010a, 2010b), and economic data from the International Monetary Fund (IMF 2010).

Table 5.2 Comparison of per capita urban final energy (GJ/capita), GRP or GDP (Int. $2005/capita) and energy intensity (MJ/Int.$) statistics (number of observations and indicator values) at the urban compared to national levels. Data cover 225 urban areas, of which 160 were located in Annex-I countries. Reported data refer approximately to the year 2000, albeit different city studies report different base years. Average and standard deviations (SD, in italics) are presented for three sample groups: 'lower', all those cities in which urban indicators are below the respective national averages; 'higher', indicators are higher than the national average; and 'Total', indicators for all cities in the sample taken together

| | | count (nr of urban areas) | | | GJ/cap | | International 2005 $/cap | | MJ final energy/ International 2005 $/cap | |
		per capita urban final energy, compared to national	per capita urban GRP, compared to national GDP	final energy intensity of urban GRP, compared to national GDP	average per capita urban final energy	SD per capita urban final energy	average per capita urban GRP	SD per capita urban GRP	average energy intensity urban GRP	SD energy intensity urban GRP
global	lower	151	111	141	88	33	17,881	12,323	3.5	3.2
	higher	74	114	84	134	105	30,265	13,219	10.6	10.2
	total	225	225	225	103	69	26,192	14,160	5.8	7.2
UNFCC Annex I countries	lower	129	82	116	96	25	29,694	9,293	3.0	0.9
	higher	31	78	44	187	128	33,639	10,681	6.8	4.5
	total	160	160	160	114	70	32,875	10,515	3.8	2.6
Non-Annex I countries	lower	22	29	25	38	30	9,364	4,999	5.9	7.8
	higher	43	36	40	95	61	10,479	8,527	13.4	12.2
	total	65	65	65	76	59	9,742	6,367	10.9	11.4
Non OECD Asia*	lower	6	14	11	55	54	9,938	4,905	4.0	1.9
	higher	37	29	32	88	38	14,538	14,112	10.1	5.6
	total	43	43	43	83	41	10,580	6,851	9.3	5.6
Latin America, Middle East and Africa	lower	16	15	14	32	12	5,826	4,383	6.7	9.0
	higher	6	7	8	141	134	8,957	5,133	33.8	21.2
	total	22	22	22	62	83	8,103	5,043	14.1	17.8
OECD90	lower	122	80	104	97	25	32,857	6,631	3.0	0.9
	higher	25	67	43	208	133	34,365	10,471	6.6	4.3
	total	147	147	147	116	72	34,108	9,921	3.6	2.4
REF	lower	7	2	12	73	22	16,516	6,979	3.5	0.6
	higher	6	11	1	98	43	20,993	5,012	7.6	5.6
	total	13	13	13	85	35	18,927	6,185	5.4	4.3

* includes in this aggregation for pragmatic reasons also Belgrade, which in fact is non Annex-I Europe

5.2.3.2 Per capita energy use

An initial observation is that almost two out of three urban areas studied have a *lower than national average (direct) final energy use on a per capita basis*. This trend is even more pronounced (in 129 of 160 cases) among Annex-I countries (which are overrepresented in this sample with 160 out of 225 cases) compared to (22 of 65 cases) among non-Annex-I countries. A primary reason for this is the effect of urban economic structures with a higher share of less energy-intensive service activities compared to national averages. An additional factor is a notably different final energy mix at the urban level in Annex-I cities, with the consumption of grid based carriers (such as gas, district heat and especially electricity) exceeding the national average. To some degree, also the effect of urban density on lower transport energy use (more public transport and soft mobility modes compared to national averages that reflect rural automobile dependence) are reflected in this pattern, but it should be noted that in Annex-I countries the non-urban population is typically a relatively smaller share of the total.

In non-Annex-I urban areas the reverse pattern is observed, with more than two out of three urban areas having *higher per capita final energy use* compared to their respective national averages. Among non-Annex-I countries although, there is pronounced *regional heterogeneity*: Africa and Latin America share the prevalence of *lower than national average urban per capita final energy use* of Annex-I countries, in contrast to Asia where urban per capita final energy use is predominantly (37 out of 43 cases) *higher than the national average*. These patterns reflect primarily the much higher urban incomes in Asia compared to rural areas and the national average leading to a much higher final energy use as well. In many Asian cities of our sample, particularly in China, urban energy infrastructure deficits are also less pronounced thus being unlikely to effectively limit energy demand. Several of the rapidly growing African cities studied in contrast, are facing acute shortages in access to modern energy carriers for the rapidly growing urban areas.

To explain these differences requires further analysis, but preliminary findings suggest that differences in levels of incomes and in economic structure (degree of service versus industry orientation of urban economies) are likely to be the main explanatory variables. In general, the number of observations in rapidly growing economies of non-OECD Asia is much larger in the sample than those of Latin America and Africa (43 versus 22), illustrating the need for improved energy information in urban settlements in these regions particularly.

Figures 5.1 and 5.2 summarize this statistical analysis, showing all the city observations as a cumulative plot (over population) sorted by decreasing per capita final energy use. The color code indicates whether a city is above (red) or below (blue) its respective national average (see color plates 6 and 7). The inverse per capita energy use pattern of cities in Annex-I versus non-Annex-I countries is clear from this comparison. On average, the *lower energy-use cities in Annex-I countries have a final energy use that is one-third lower than the Annex-I national average*. For non-Annex-I countries the relationship is inverse: *most non-Annex-I cities have higher (about twice) per capita final energy use* than their respective national averages, being in the same ballpark as the lower energy use city sample in the Annex-I countries (at some 100 GJ/capita).

These conclusions only refer to the (direct) final energy use metric adopted for the comparative analysis of our sample of 225 urban areas.

Evidence suggests that for Annex-I country cities the lower final energy use is likely to hold only for the production-accounting approach adopted for this comparison. Adding 'embodied' energy use (corrected for net trade of imports and exports of manufactured goods and services from and to urban economies) and adding upstream energy sector conversion losses is likely to weaken the conclusion of a lower urban energy footprint in cities of Annex-I countries compared to the national average (see Chapter 4) as lower (direct) final urban energy use is likely to be (largely) compensated by higher 'embodied energy' consumption associated with higher urban incomes. And yet, the lower (direct) final energy use of many urban compared to rural areas in Annex-I countries illustrates well an energy-related *urban comparative advantage*: urban areas with their corresponding more energy-efficient compact settlement structures and lesser (energy-intensive) automobile dependence and greater potential for efficiency improvements through energy system integration ('recycling', i.e., co-generation and heat cascading) have larger efficiency-leverage potentials compared to those of rural areas. The challenge to reduce the energy and environmental footprint from (over)consumption (i.e., embodied energy) is not unique to urban dwellers as it applies equally to rural ones in Annex-I countries since urban–rural income differences there are typically less pronounced.

The situation of cities in non-Annex-I countries, particularly in Asia, is markedly different. Compared to rural areas, cities not only have higher (direct) final energy use, they also have generally much higher incomes. Thus, the urban–rural gradient in terms of per capita (direct) final energy use is amplified yet more by higher urban incomes, which further increases the rural–urban energy gradient when considering the 'embodied' energy use associated with consumption. Given the dynamics of urbanization trends (see Chapter 2), it is thus fair to conclude that the urban energy demand 'hot spot' in the decades to come will reside particularly in the rapidly growing cities of non-Annex-I countries, especially in Asia.

5.2.3.3 Per capita income

Regarding per capita income the data sample reveals much more heterogeneity than popular conceptions of invariably rich urbanites would suggest.[6] Almost half of the urban areas in our sample had per capita GRP/GDP values below the national average. Again, the patterns diverge between Annex-I countries (where this trend is driven by the large number of relatively deprived smaller UK urban centers in the data sample, but also by such prominent examples such as the capital of Germany, Berlin) and non-Annex-I countries.

In non-Annex-I urban areas the majority showed above national average per capita GRP/GDP values. In Asia, two out of three urban areas had GRP/GDP above the national per capita average. In Africa and Latin America, just over one-third of the urban areas had GRP/GDP values that exceeded the national per capita average, but two-thirds ranked below it.

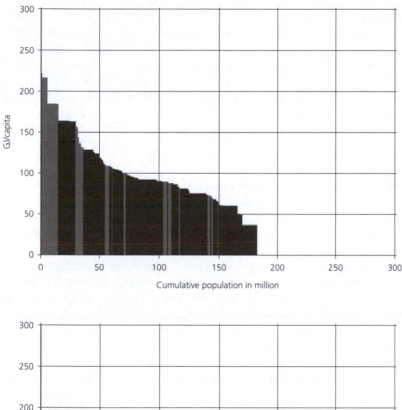

Figure 5.1 Per capita (direct) final energy consumption (TFC) (GJ) versus cumulative population (millions) in urban areas (*n*=160) of Annex-I (industrialized) countries. Red indicates urban areas with per capita TFC *above* the national average. Blue indicates per capita TFC *below* the national average. (See color plate 6)

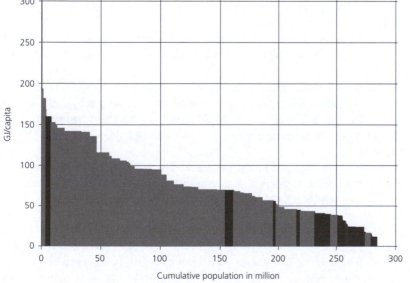

Figure 5.2 Per capita (direct) final energy consumption (TFC) (GJ) versus cumulative population (millions) in urban areas (*n*=65) of non-Annex-I (developing) countries. Red indicates urban areas with per capita TFC *above* the national average. Blue indicates per capita TFC *below* the national average. (See color plate 7)

5.2.3.4 Energy intensity

Regarding energy intensity of urban GRP/GDP, the majority of urban settlements studied showed lower than national average energy intensities, which indicates the dominance of less energy-intensive tertiary sector activities in most urban areas. In Annex-I countries more than two out of three settlements show lower than national level energy intensities of GDP. In the non-Annex-I countries the general trend is almost balanced, with just a few more cases of urban energy intensity that exceed the national average values. Again, the Asian urban areas show a very different pattern to those from the Latin America, Middle East, and Africa regions. Three out of four urban areas in non-OECD Asia have energy intensities that exceed the national average, which points to the rapid (to a good extent also export-oriented) industrialization as driver of urbanization dynamics in those countries, while four out of five of the African, Middle East, and Latin American urban cases have the same pattern as OECD countries, with urban area energy intensities below their respective national average.

5.2.3.5 General observations

For the non-Annex-I urban areas at least three general patterns of energy use can be discerned.

One is the lower end, with final energy use under 30 GJ/capita. Per capita income is mostly fewer than 5,000 Int. $/capita (in PPP terms) and energy intensity is, in many cases, also quite low, below 5MJ/$. This low energy intensity in low-income, non-Annex-I cities does not necessarily suggest highly efficient energy systems, but rather different consumption structures (particularly lower private transport energy use). In all likelihood, the low energy intensities may also reflect an underreporting of noncommercial, traditional biofuels used by low-income urban households.

The medium range of per capita final energy use in non-Annex-I cities is from 30 to 100 GJ/capita and coincides with a wide range of incomes and energy intensities.

Heavy industrial urban areas show yet higher per capita final energy use of up to 350 GJ/capita and over a highly variable range of income levels. In practically all urban settlements of the third group of non-Annex-I countries energy intensity is above 10 MJ/$ (up to 39 MJ/$ in the sample).

Patterns for Annex-I cities are markedly different. The Annex-I city panel in general appears more coherent in final energy use patterns. First, the correlation between higher urban incomes and higher final energy use tends to weaken significantly, with richer cities not necessarily using more (direct) final energy on a per capita basis (but highly likely to use much more embodied energy compared to poorer cities). Second, there is a strong and inverse correlation between urban incomes and energy intensity, with the latter falling with rising urban incomes. Only seven out of the 160 Annex-I urban areas show energy-intensity values above 10 MJ/$ and the vast majority are below 5 MJ/$.

5.2.3.6 Variable correlations

Figure 5.3 presents the overall positive correlation between per capita incomes and (direct) final energy use. The general positive correlation, familiar from national level analysis, is also found at the urban scale. However, the standard derivation of energy intensity at the urban level exceeds the variation at the national level by a factor of three, which suggests a much broader spread for path-dependent urban development trajectories. Also, within the panel of Annex-I countries there is a large variation in energy consumption per capita (with some of the small industrialized urban areas using more than 600 GJ/capita final energy). A 'turning' (or saturation) point proposed in the literature (World Bank 1992; Stern 2004) cannot be identified at a statistical significant level in this data set, which covers GRP/GDP ranges up to about 80,000 Int. $ 2005/capita (in PPP terms), despite a visible weakening of the income–energy use link for high-income cities.

Figure 5.4 presents the relation of GRP per capita and energy intensity of GRP. Trends in non-OECD Asia cities come closest to the often proposed 'hill' (Goldemberg 1991), with a peak in energy intensity at about 10,000PPP$/capita and a pronounced decline in energy intensity at higher per capita incomes. At lower income ranges, however, the data in our sample are sparse, and often also exclude the dominant noncommercial traditional biofuels, so the above findings are consistent with the observation of a 'hill' in the development of *commercial* energy intensity (Goldemberg 1991) against a background of continuously falling *total* (including noncommercial) energy intensities with rising incomes (see Nakicenovic et al. 1998) as evident in the cities of Africa and Latin America in Figure 5.4.

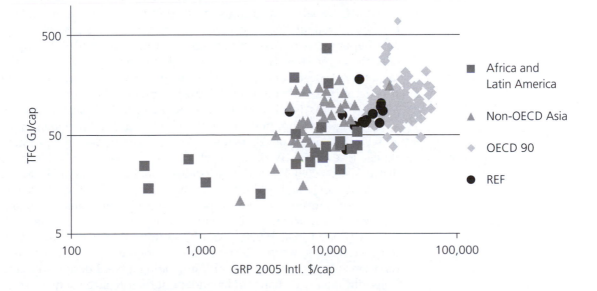

Figure 5.3 Comparison of urban total final energy consumption (TFC in GJ) and urban income (GRP/GDP at PPP in Int. $2005) per capita for cities in Annex-I and non-Annex-I countries

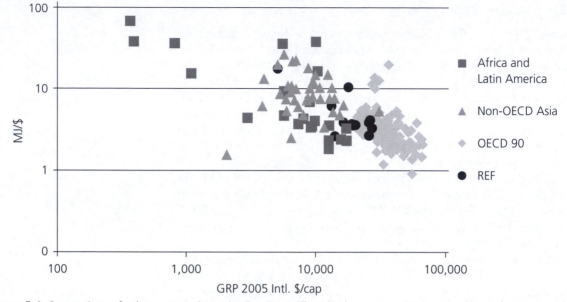

Figure 5.4 Comparison of urban energy intensity (in MJ/Int. $) and urban per capita income (in Int. $ 2005 at purchasing power parities) for cities in Annex-I and non-Annex-I countries

5.2.3.7 Rank–size distribution patterns

Given the significance of urban areas for energy use and global resource consumption and the rapidly increasing urbanization trends, an improved understanding of their size-distribution and likely growth dynamics will be crucial for the management of a successful transformation towards more sustainable futures. For many physical, biological, social, and technological systems for example, robust nontrivial quantitative regularities like stable patterns of rank-distributions have been described (see also Chapters 2 and 3). Examples of such power-law-scaling patterns include phenomena such as the frequency of vocabulary in languages, the hierarchy of urban population sizes across the world (Berry et al. 1958; Krugman 1996) or the allometric scaling patterns in biology, such as Kleiber's law (which notes the astonishing constancy in the relation of body mass and metabolic rates in living organisms across many orders of magnitude). In ecosystem studies (West et al. 1999; Brown et al 2008) as well as in architecture (Batty 2005, 2008) there is vivid debate about the extent to which underlying constraints of hierarchical branching networks of metabolic systems or transport network capacities are ultimate causes defining the size, shape, and rank-distribution of (organic or urban) network entities (Decker et al. 2000, 2007). Questions that derive from such consideration are, whether all cities are (analogous to e.g. all mammals) variations of an identical blueprint, inhabiting a basal design structure of specific energy demand, independent of their realized actual size or 'current incarnation'? To find out if such scale-free regularity applies to the urban cases collected for this assessment, the 225 data points of the

GEA City energy database were ranked by total final energy consumption as displayed in Figure 5.5 which stretches from 0.825 PJ for the city of Shinayanga in Tanzania with about 50,000 inhabitants to 2 EJ for the 14.5 million residents of the Shanghai metropolitan area.

The analysis reveals significant deviations from a uniform scaling trend as proposed in a number of publications (Bettencourt et al. 2007), with a distribution characterized rather by threshold effects across an overall convex distribution. If the whole data set is subdivided into three individual panels (differentiated by energetic city size) significant differences in exponents were found, suggesting important threshold effects: small cities (below 30 PJ of final energy use) presented the steepest incline in energy use with increasing city rank (i.e. size): in doubling their rank position, cities of those consumption range tend to increase their energy use by a factor of 6.1. For the medium-sized panel of urban energy use cases, which used between 30 and 500 PJ, a doubling of city rank corresponds to an increase in energy consumption only by the factor of 1.6. In the panel of the largest urban areas in the data set, which all used more than 500 PJ final energy per year, a doubling of urban rank is associated with an amplification of urban energy use by a factor of 0.5 only. This indicates considerable positive agglomeration economies of bigger cities with respect to energy use. Only four urban areas of the entire sample of 225 cities have in fact an annual final energy use that is significantly above one EJ. On the other hand urbanization currently is most rapid in the smallest segment of city sizes (below 500,000 inhabitants), which according to population projections will in the future host a steadily increasing share of the global urban population (UN DESA 2010) and based on above-observed patterns will be characterized by a very high elasticity of energy demand growth with respect to their further increases of urban size.

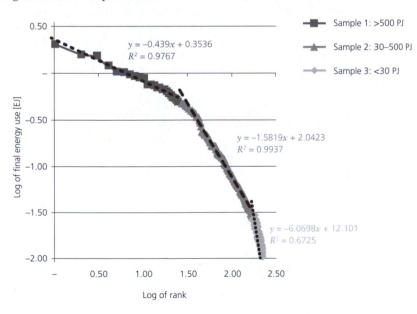

Figure 5.5 Rank–size distribution of urban energy use (three subsamples)

According to the data set presented here urban areas appear to be 'a different kind of animal', showing considerable deviations from a scale-free distribution of energy use pattern. This finding, while tentative and restricted to the data sample available for analysis, suggests a need for deepened analysis and improved modeling of urban energy use with explicit consideration of city size heterogeneity both in terms of urban growth rates as well as their energy implications.

Notes

1   The absence of appropriate spatially explicit data sets of energy conversion processes and systems (power plants, refineries, etc.) does not allow to follow a similar 'downscaling' approach for upstream conversion losses in the energy system, which would be needed to derive spatially explicit data on primary energy use. The study therefore is restricted to estimated urban energy use at the final energy use level.

2   The lack of available IEA regional estimates limits the possibilities for a more detailed comparison, but in the reported four IEA regions, urban energy use is within the respective minima/maxima regional values of the IIASA study.

3   The main source of uncertainty for the ranges reported in Table 5.1 is the fuzziness in delineating urban areas and population and hence the attribution of national energy use to the urban category (e.g. in including or excluding peri- and sub-urban areas in the energy use estimations). Conversely, the uncertainty in energy statistics is comparatively small, with the main uncertainty source being the lack of reliable data on urban noncommercial (traditional biomass) energy use, particularly in Africa.

4   Annex-I countries as defined in the UNFCCC (the United Nations Framework Convention on Climate Change).

5   Table 5.2 presents an updated and extended summary from the GEA city energy database with a sample of 225 cities (compared to a sample of 200 cities presented in the original GEA urbanization chapter).

6   GRP (gross regional product) data is provided only by a limited number of statistical offices or other sources. They differ in methodology and are not always strictly comparable. A more detailed discussion of economic measurement issues at the urban scale is beyond the scope of this energy assessment.

Part II

THE URBAN CHALLENGES

6

Energy access and housing for low-income groups in urban areas

David Satterthwaite and **Alice Sverdlik**

6.1 Introduction

As this chapter will describe, several hundred million urban dwellers in low- and middle-income nations lack access to electricity and are unable to afford clean, safe fuels. The health and other developmental advantages of access to electricity and cleaner fuels are given too little consideration, and a high proportion of low-income urban households remain dependent on polluting fuels that impose major health, cost, and time burdens. If electricity and natural gas can be accessed via legal connections, at an affordable price and reliable supply, then households will not struggle with the time burdens of purchasing or gathering fuels. No space in the home is needed to store these cleaner fuels (and most low-income urban dwellers live in overcrowded conditions). Reliable electricity supplies bring many other obvious advantages – reliable, cheap, and safe lighting at night; the possibility of fridges, televisions, and electric fans; support for home enterprises; and a very large reduction in fire risk.

In the development literature, energy is not generally recognized as one of the basic needs (Pachauri et al. 2004) and appropriate measures are rarely utilized to capture energy poverty among low-income urban households. The eight Millennium Development Goals and the 18 Millennium Development Targets include targets for reducing hunger and income poverty and for education, water, and sanitation but not for safe, clean energy. However, the issue of energy poverty is considered in high-income nations, although this is not in regard to fuel use but in regard to the substantial proportion of household income spent on fuels and electricity (typically if more than 10 percent of household income is spent on household energy) (Boardman 1993; Buzar 2006). This is not an appropriate measure for much of the urban population in low-income and some middle-income nations because their incomes are so low in relation to the costs of food and necessities other than food that their energy use is very low. This is both in the energy used within their homes (lighting, cooking, and, where relevant, space heating and appliances), in the energy implications of the transport modes they use, and, for those who are self-employed, in the energy used in their livelihoods. Rather than percent of household income, the appropriate indicator for their 'energy poverty' is their incapacity to afford energy sources and in the poor quality of the energy sources they use. Such individuals or households also have so few consumer goods that their individual embodied energy (including their carbon budget) is also low. Thus, for nations where many low-income urban households use noncommercial fuels and keep energy expenditures down by using dirty fuels (including wastes) or cutting fuel consumption, the proportion of income spent on energy is not a good indicator of poverty.

In addition, an analysis of poverty in relation to energy should also consider the time and effort used to obtain fuels, the health implications (including those that arise from indoor air pollution and the risks of fire and burns), and the quality and convenience of the fuels used to meet daily needs (i.e., in space heating or cooling, and for hot water). Pachauri et al. (2004) suggest that, ideally, the analysis of the adequacy of energy should include primary energy consumption, end-use energy (especially electricity), useful energy (e.g., whether the primary or end-use energy delivers the energy needed), and the quality and adequacy of energy services for households (including transport). However, data are often only available for the first two of these. Moving out of poverty involves shifts away from the dirtier and less convenient fuels[1] and obtaining access to electricity, as well as keeping down total monetary expenditures on energy. Thus, the two most common implications of poverty in regard to energy use among urban populations in low- and most middle-income nations are, first, use of the cheapest fuels and energy-using equipment (including stoves, which bring disadvantages, especially in regard to indoor air pollution, inefficient fuel combustion, and convenience) and, second, no access to electricity.

Greater attention is needed to the implications of energy poverty for safety, health, well-being, as well as households' coping strategies under conditions of energy scarcity. But data are often lacking on these dimensions. Low-income households may limit the number of meals (in extreme circumstances to one a day) to save money both on food and cooking fuel. Poverty is also evident in the lack of space heating within cold climates or seasons – although this is difficult to measure as expenditure surveys cannot identify what consumers forgo. Lack of electricity also limits many income-earning possibilities for home-based enterprises. Urban poor households often face much higher risks of burns and scalds for household members (especially children) and of accidental fires, underpinned by a combination of extreme overcrowding (often three or more people to each room), unsafe fires or stoves, the absence of electricity for lighting (candles and kerosene lights are used), housing built of flammable materials, high-density settlements, a lack of firebreaks, and no emergency services, including fire services (Hardoy et al. 2001; Pelling and Wisner 2009). All of the above are also often associated with homes and livelihoods in informal settlements – which helps explain the lack of electricity (with utilities or companies unwilling or not allowed to operate there), the poor-quality housing, and the lack of provision for fire-prevention and emergency services in most such settlements.

Discussions of energy poverty should also consider the cost burdens experienced by low-income households who have access to electricity and who use cleaner, more convenient fuels. For instance, in cities such as Mumbai (India), low-income households who move from informal to formal housing frequently benefit from access to electricity, but often find it difficult to pay the bills. A high proportion of low-income urban dwellers rent accommodation and landlords may be charging them more for electricity than they pay. A survey of 300 homes in Rio de Janeiro's informal settlements found that liquefied petroleum gas (LPG) was the main cooking fuel, while electricity was used for lighting and appliances, so the problem was not so much the quality of energy but the fact that households were spending 15 to 25 percent of their incomes on energy (WEC 2006). Here, there are more parallels with energy poverty in high-income nations. Buzar (2006) noted that increasing numbers of households in former Communist states in Eastern and Central Europe[2] were facing difficulties in affording energy, in part because of significant energy-price increases as subsidies were removed, and in part because of the failure of the state to develop safety nets to protect low-income groups. This leaves many families with no option but to cut back on energy purchases, a problem further aggravated by cold climates and the lack of insulation and other aspects of the low energy efficiency of the building stock.

6.2 Housing quality and location

Around 800 million urban dwellers in low- and middle-income nations live in poor-quality, overcrowded housing with inadequate provision for basic services (UN HABITAT 2003, 2008). A taxonomy of their housing submarkets with associated energy implications is given in Table 6.1.

Low-income groups in urban areas face limited choices in renting, buying, or building accommodation that they can afford and so have to make trade-offs between a good location (especially in regard to income-earning opportunities), housing size and quality, infrastructure and service provision, and secure tenure (see references for Table 6.1). Good locations in relation to income-earning opportunities mean that transport expenditures and time can be kept down, and more central locations usually have greater possibilities of infrastructure and service provision. But they are also more expensive because of competition for space, so that residents often seek to reduce their housing costs through illegally occupying land or purchasing an illegal subdivision and self-built homes. At their most extreme, to obtain central locations, low-income groups live in shacks built on pavements or waste dumps; alternatively, they may occupy small rooms with more than three people to a room or share beds (so a single person pays to sleep in a bed in a shared dormitory with each bed serving two or three people over a twenty-four-hour period, known as 'hot beds').

One of the most extreme examples of this are the tens of thousands of pavement dwellers in Mumbai, where the choice to live on the pavement (and usually with low lean-to shacks too small to sleep in) results from a combination of their very low incomes, the central location of where they earn their incomes (they walk to work), and the impossibility of affording transport costs from less central locations (SPARC 1990). Another example are households in Indore (India) who choose to live on land sites adjacent to small rivers that flood regularly. These have economic advantages because they are close to jobs or to markets for the goods the households produce or collect (many earn a living collecting waste). The land is cheap and, because it is public land, the residents are less likely to be evicted. These sites have social advantages because they are close to health services, schools, electricity, and water, and there are strong family, kinship, and community ties with other inhabitants (Stephens et al. 1996). In Dhaka, the large informal settlement called Korail may be overcrowded with poor-quality housing and inadequate provision for infrastructure and services but for low-income households, it is close to income-earning opportunities (Jabeen et al. 2010).

Table 6.1 The housing submarkets used by low-income urban dwellers and their energy-use implications

Housing type	Energy implications in the home	Energy implications for transport
Rooms rented in tenements	Typically one room per household or one room shared by several individuals. Often electricity available, but usually too expensive to use for cooking and space heating	Usually close to sources of livelihood or demand for casual work (hence this type of accommodation is in demand even with high levels of overcrowding)
Cheap boarding houses/ dormitories (including hot beds)	Very low energy use; no provision for cooking?	As above
Informal settlement 1: squatter settlements (in many cities these house 30–60% of the entire population)	In low-income nations, usually reliance on dirtier fuels and lack of electricity; in many middle-income nations less so; for many households, part of fuel/electricity expenditure is for livelihoods in the home; illegal electricity connections may be common; often high risks from accidental fires	Many in peripheral locations, which implies high transport costs in time and money; better located squatter settlements often become expensive through informal rental or sale
Informal settlement 2: housing in illegal subdivision (which has not received official approval)	The land is more expensive than illegal land occupation, but less at risk from eviction and often with more provision of infrastructure (including electricity) or at least more possibilities of provision as the land occupation is not illegal	Many in peripheral locations which implies high transport costs; in large cities, the cheapest illegal subdivisions can imply several hours traveling a day to and from sources of income
Accommodation at the workplace	Common for single men in some cities; extent not known and includes apprentices	
Pavement dwellers and those who sleep in open or public spaces	Very low incomes so very low fuel use	Walk to work

Sources: Hardoy et al. (1989); Yapi-Diahou (1995); Harms (1997); Mitlin (1997); Mwangi (1997); Bhan (2009)

In more peripheral locations, rented accommodation or land on which houses can be built is cheaper but households' transport costs are higher and access to work may be sharply curtailed. If peripheral land can be occupied with no payment, it is usually more distant from income-earning opportunities. This means high time- and monetary-transport costs, and it is difficult to establish such elevated transport costs for those living in peripheral locations because most of the data on the proportion of income spent on transport are simply urban averages. In addition, it is likely that many household surveys under-represent those who live in informal settlements – for instance, a lack of formal addresses and maps makes it difficult to include their inhabitants in surveys or those responsible for collecting data fear to work in informal settlements (for an example of this from Cairo, see Sabry (2009)). Peripheral locations also constrain the inhabitants' access to economic opportunities, as many locations are too distant or too expensive to commute to.

6.3 Urban populations and energy use in low- and middle-income nations

In most urban centers, existing data are unable to capture the key dimensions of energy poverty reviewed above. There are some general statistics on the forms of energy use for urban populations – for instance, in what fuels (and mix of fuels) are used and whether or not they have access to electricity (Table 6.2). However, there are no general statistics on how fuel use and access to electricity vary within nations' urban populations or within cities by income group. In part, this is because many 'energy' statistics for individuals or households are only available for national populations from sample surveys. Where these can be disaggregated, it is often only as averages for 'urban populations' so this obscures large differences between different urban centers and between different income-groups within each urban center. In part, this is because the documentation of 'energy' provision deficiencies has not been given the same level of attention as, say, deficiencies in provision for water and sanitation.[3] The only exception is the very considerable documentation on the health impacts of pollution from the use of 'dirty' fuels (and other factors, including poor ventilation and inefficient stoves), although much of this literature is for rural households and perhaps underestimates the extent of this problem among low-income urban households (WHO 2006).

Table 6.2 shows how most (90 percent) of the urban population in 'developing countries' had access to electricity and 70 percent had access to modern fuels (mostly gas) in 2007 – but also how the picture on energy access for urban populations was very different for the least-developed countries and for sub-Saharan Africa. If 30 percent of the urban population among low- and middle-income nations lack access to modern fuels, this implies a total figure of nearly 700 million urban dwellers. A higher proportion has access to electricity – but about half the urban population within the least-developed nations and

Table 6.2 The proportion and number of the urban population that lacks electricity and access to 'modern fuels'[4] in developing countries, least-developed countries, and sub-Saharan Africa

Percentage and number of the urban population	Developing countries	Least-developed countries	Sub-Saharan Africa
Lacking access to electricity	10% (226 million)	56% (116 million)	46% (124 million)
Lacking access to modern fuels	30% (679 million)	63% (130 million)	58% (156 million)

Sources: UNDP and WHO (2009).[5] Statistics on the urban population are drawn from UN Population Division (UN 2008) and are for 2005.The dates for the statistics on access to electricity and modern fuels vary by country, with most being between 2003–7

within sub-Saharan Africa lack access to electricity. In sub-Saharan Africa alone, this implies that around 120 million urban dwellers lack access to electricity. For all low- and middle-income countries taken together, some 230 million urban residents lack such access.

Particular studies suggest that it is common for low-income urban households in Africa and Asia to use a mix of fuels – for instance, different fuels for different kinds of food and fuel-switching at certain times of year when fuel prices or household incomes change (for China, see Pachauri and Jiang (2008); for Arusha, see Meikle and North (2005)). Policymakers rarely take into account these complexities. Regional and seasonal differences may be significant, and households are also influenced by subsidies and incentives, fuel availability, and cultural preferences. For instance, in a study of energy use in an informal settlement in the Cape Peninsula in South Africa, most households with legal electricity connections and meters could access a tariff with 50 kw hours a month free basic electricity which encouraged them to cook with electricity rather than paraffin (Cowan 2008).

Energy use in low-income households thus varies widely and may change over time, and these patterns may differ significantly from higher-income urban residents with access to cleaner fuels. Energy use by low-income urban dwellers ranges from very large numbers who use little or no fossil fuels and electricity (i.e., wood, dung, straw, charcoal, and biomass wastes) through greater use of solid and liquid fuels (kerosene and/or coal or coal-based fuels, often called transition fuels) and a proportion of households with electricity to the use of cleaner fuels (bottled or piped gas) and electricity. Available studies also give examples of the often large differentials in energy used between high-income and low-income households within particular urban centers; some show that these can vary by a factor of ten or more, but of course the scale of the differentials depends, in part, on how 'the urban population' is divided for this comparison (e.g., differentials will be greater if the richest and poorest deciles are compared instead of the richest and poorest quartiles).

The two nations with the world's largest urban populations are China and India; by 2010 these accounted for more than a quarter of the world's urban population (UN 2009). In India, fossil-fuel-based energy sources increasingly dominate the energy mix of urban households, although biomass (including firewood and dung) continues to be used, especially by the lowest income groups. Between 1983 and 2004/5, there was a rapid rise in the use of LPG and electricity among urban households (Pachauri and Jiang 2008). The percentage using LPG grew from 9 to 61 percent while the percentage with electricity rose from 58 to 91 percent. The proportion of the urban population using fuel wood fell from 61 to 35 percent with those using dung declining from 27 to 10 percent and coal/coke from 21 to 5 percent (Pachauri and Jiang 2008). Data on cooking fuels used by India's urban populations from the 2005/6 demographic and health survey give a comparable picture – with 59 percent using gas or LPG, 23 percent using firewood or straw and 8 percent using kerosene; see Table 6.4). India's national surveys found that in 2000, 7–15 percent of the bottom two urban deciles used LPG versus nearly 70 percent in the top two deciles (Gangopadhyay et al. 2005).

In China, among urban households there has been a shift away from the direct use of coal to gas and electricity (although coal is still important for a significant proportion of urban households). Energy consumption among urban households declined from 1985 to 2002 (from 9 GJ/capita to around 5 GJ/capita), because of a shift to more efficient fuels (Pachauri and Jiang 2008). However, dependence on coal may not be reduced if coal-fired power stations are an important part of meeting the resulting rising demand for electricity.

The diversity in the forms of energy used by low-income urban dwellers between nations is mirrored in variations within and across a nation's cities, so that any generalizations about energy use among low-income urban dwellers must be treated with caution. The only obvious characteristics that such households share is limited purchasing power for energy (for all uses) and a desire to keep costs down, so their fuel use and fuel–energy mix is much influenced by the price and availability of different fuels and electricity. Having access to electricity at prices that low-income households can afford obviously represents a major advantage – for lighting and for key appliances (including fridges and, where needed, fans), for household enterprises and for the reduced risk of accidental fires. However, they will keep electricity consumption down (for instance, where it is expensive in comparison to other fuels, they may not use it for cooking or space heating) unless there is no better alternative (or they have made illegal connections to power grids that keep costs down).[6] Having gas for cooking and hot water (and, where needed, space heating) has great advantages of convenience and of low generation of indoor air pollution, but in many urban contexts it is only available as LPG canisters (and so less convenient and more expensive than gas piped to the home). This often makes it too expensive for large sections of the urban population.

Among the low-income households in urban centers in the lowest-income nations, fuel use is dominated by charcoal, firewood, or organic wastes (e.g., dung). As access to fuels is commercialized, less fuel can be afforded or households shift to cheaper (usually dirtier) fuels. In many small urban centers in low-income nations, it may be that certain fuels (wood, dung, agricultural wastes, etc.) are cheap and that a proportion of the urban population can gather fuel rather than pay for it – but probably the larger the city, the greater the commercialization of all fuels. Also, the very limited space within the homes of most low-income urban households – especially those that live in central areas with, in some cases, less than 1 m²/person – limits the capacity to store bulky solid fuels.

It is also common for low-income households to operate a small-scale business out of their homes, and energy may be a key part of the business – and a major expense. For instance, a survey of two neighborhoods in Salvador (Brazil) found that 57 percent and 96 percent of enterprises were located in the owner's house (Winrock International 2005). Energy consumption among these entrepreneurs, who were often women, could reach 30 percent of household income (Alvarez 2006). Many women extended their domestic tasks like sewing or cooking to include income-earning, which required energy – for instance electricity for sewing machines and for food preparation, refrigeration, and lighting. Productive and household energy requirements are thus interlinked among many low-income urban households, and affordable energy is important to the success of home-based enterprises.

Energy requirements were sizable for small-scale entrepreneurs in Kibera, one of Nairobi's largest informal settlements, and many of these are home-based businesses. Electricity use was limited in the home-based enterprises, due to the large upfront connection costs (Karekezi et al. 2008). Over 42 percent of the enterprises used charcoal, while electricity utilization stood at 29 percent, fuelwood at 19 percent, and kerosene at 9 percent. Access to electricity would offer more efficient and cleaner energy for many kinds of small businesses, rather than fuels that are inefficient, polluting, and a major drain on their incomes.

6.4 Fuel use for cooking

Table 6.3 shows the contrast between the proportion of the urban population using wood, charcoal, and dung for cooking in developing countries (less than one-fifth of households), in the least developed nations (two-thirds of urban households), and in sub-Saharan Africa (more than a half). In developing countries close to two-thirds of the urban population use gas or electricity for cooking; for the least-developed nations and sub-Saharan Africa, this is less than one-quarter. There are large differences in this within the least-developed nations and in sub-Saharan Africa. For instance, for many of these nations only a small percentage of the urban population has access to electricity.

Table 6.4 gives details of type of cooking fuel used by urban populations and access to electricity in forty-three nations, drawing on data from the Demographic and Health Surveys. Among the nations for which data were available, 20 had more than half of the urban population relying on non-fossil fuel cooking fuels – charcoal, fuelwood, straw and dung.

Table 6.3 The main fuels used for cooking in urban areas in developing countries, least-developed countries, and sub-Saharan Africa (% of urban population using particular fuels)

Percentage of the urban population	Developing countries	Least-developed countries	Sub-Saharan Africa
Using wood, charcoal, and dung for cooking fuels	18	68	54
Using coal for cooking	8	3	2
Using kerosene for cooking	7	4	20
Using gas for cooking	57	20	11
Using electricity for cooking	6	4	11

Sources: UNDP and WHO 2009

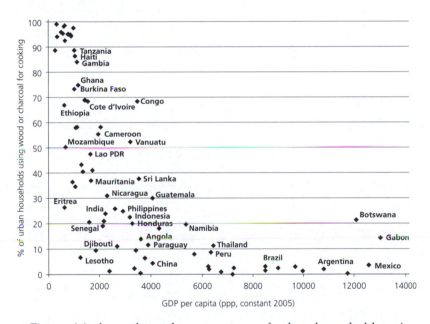

Figure 6.1 Nations' per capita GDP against percentage of urban population using wood/charcoal for cooking
Sources: Data on wood and charcoal use for cooking drawn from UNDP and WHO (2009); data on GDP per capita from World Development Indicators

Figure 6.1 shows how the percentage of urban households using wood or charcoal for cooking falls with a nation's per capita GDP, although it also shows the considerable variation in this among lower-income groups with comparable per capita GDPs.

In Figure 6.1, for most nations with per capita GDPs under $1,100, 85 percent or more of their urban population use wood and charcoal for cooking – and all these nations are in sub-Saharan Africa, except Haiti. For nations with per capita GDPs above $14,000 virtually all urban households do not use wood or charcoal. For nations with per capita GDPs of $1,100–4,000, the variations in the percentage of the urban population that use wood or charcoal are very large (UNDP and WHO 2009).

Households select fuels for food preparation for reasons that include cost, availability, convenience, type of food, and cooking equipment, as illustrated by a study in Ibadan (Nigeria). Kerosene

Table 6.4 Cooking fuel and access to electricity for urban populations

Nation and year of survey	Has electricity			Type of cooking fuel								
	No	Yes	Missing	Electricity	LPG & natural gas	Biogas	Kerosene	Coallignite	Charcoal	Firewood, straw	Dung	No cooking in household and 'other'
Armenia 2005	0	99.9	0.1	15.9	39.9	43.4	0.1	0	0	0.6	0	0
Azerbaijan 2006	0.2	99.8	0.1	14.9	76.3	7.7	0.1	0	0	0.8		0
Bangladesh 2004	23.4	76.6	0	0.3	30.9	0.5	1.3	0	0.1	40.2	5.5	21.1
Benin 2006	43.4	56.6		0.1	9		3.2		43	43.3		1.2
Bolivia 2003	5.8	94	0.2	0.8	88.2		0.4			7.3	0.2	3.1
Burkina Faso 2003	47.5	52.4	0.1	0	17.5	0.3	0.7	1.2	15.4	59.9	0.1	4.9
Cambodia 2005	33.1	66.8	0	0.6	30.1	0.1	0	25.4	43.6	0		0.2
Cameroon 2004	22.9	77.1	0		25.5		13.5		4.6	51.9		4.3
Chad 2004	83.3	16.4	0.3									
Colombia 2005	0.7	99.3		8.3	85.7		1	0.2		2.5		2.4
Congo (Brazzaville) 2005	49.2	50.8	0	4.9	15.6	8.9		49.6	18.7	1.2		0.6
Congo Democratic Republic 2007	63.2	36.6	0.2	10.9			0.3		52	36		0.8
Dominican Republic 2007	1.4	98.6		0	92.1				1.9	1.2		4.6
Egypt 2005	0.2	99.8	0	0.2	98.5	0	1	0	0	0	0	0.1
Ethiopia 2005	14.3	85.7	0	1	0.9	0.3	25.9	0.7	18.1	48.7	2.1	3
Ghana 2003	23.1	76.9	0	0.6	14.6	0.7	1.3		54.1	25.6	0	2.3
Guinea 2005	35.5	63.8	0.7	1.1	0.3	0	0.2	59.6	33.8	0.8		3.5
Haiti 2005/06	31.1	68.9	0	0.1	4.6	2	6	0.3	76.5	10		0.5
Honduras 2005				33.6	32.6		10.4	0.1	0	20.1		3.2
India 2005/2006	6.9	93.1	0	0.9	58.7	0.5	8.2	4.3	0.5	23.3	2.8	0.8
Indonesia 2002/2003	1.9	98.1	0.1	0.7	18.6		63.8	0.1	0.1	15.9		0.8

Jordan 2007	1	99		0.2	99.6		0.1	0.2	0		0	0.1
Kenya 2003	49.8	50.2	0	1	10.8	0.3	50.8	0	25.9	9.4	0.5	1.5
Lesotho 2004	73.6	26.2	0.1	7	58.2				0	6.6		27.6
Liberia 2007	92.9	6.9	0.2									
Madagascar 2003/2004	47.3	52.7	0.1	0.9	2.7	0.3	0.2	0.7	59.4	35.5	0.1	0.1
Malawi 2004	69.6	30.2	0.2	10.6	0.1	0.1	0.2	0	41.4	47.1	0	0.2
Mali 2006	52.5	47.4	0	0.1	1.3				40	55.1	1.1	2.2
Moldova 2005	0.5	99.4	0.1	5.3	18.3	74.8	0.2		0	0	0.8	0.5
Morocco 2003/2004 A6	5.3	94.6	0.1	0.3	0	0	0	0.1	0.1	98.9		0.1
Mozambique 2003	74.9	25	0.1	2.1	4.9		1.6	40.8	1	49.3	0	0.2
Namibia 2006/2007	22.4	77.6	0	67.1	9.5	2	0.4	0	0.1	15.5	0	5
Nepal 2006	9.9	90.1		0.4	40.3	3.3	15.8	0	0.1	36.4	2.5	0.6
Niger 2006	52.7	47.2	0.1	0.6	3.4			10.4	84.9	0.5	0.1	0.1
Nigeria 2003	15	84.9	0	0.5	1.5	0.6	53.4	0.2	0.7	41.1	0.1	1.8
Pakistan 2006/07	1.5	98.3	0.1	0.2	76.1	1	0		0.1	20.1	2	0.1
Philippines 2003	7.9	92	0.1									
Rwanda 2005	74.7	25.1	0.2	0.3	0.1	0.1	0.3	1.6	37	58.2	0	2.2
Senegal 2005	19.6	80.4	0	0.4	75.9				8.9	11.5	0.1	3
Swaziland 2006	36.6	63.4		41	29.8			0.8	0.9	11.4	1.2	14.9
Tanzania 2004	61	38.9	0.1	0.9	0.4		9.6		59.2	27.8	0	2
Ukraine 2007	0.1	99.9	0	9.9	87.5	0.1	0	1.8	0.1	0.4	0.1	0
Zambia 2007	52.2	47.8	0	38.5	0	0	0	0.6	53.1	7.5	0.2	
Zimbabwe 2005/06	8.6	91.4	0.1	87.9	0	0	0.6	0	0	11.3	0	

Sources: Macro International, Inc., 2009; MEASURE DHS STATcompiler (online at: www.measuredhs.com; last accessed: June 11, 2009)

was the major cooking fuel for low- and middle-income households until subsidies on petroleum products were withdrawn in 1986. As a result of the increased kerosene and cooking-gas prices, surveyed households in 1993 had begun to use fuelwood, sawdust, and other cheaper energy sources. A follow-up in 1999 discovered that households had switched back to kerosene, while also reducing the frequency of cooking, eating cold leftovers, and substituting less nutritious but faster-cooking foods (Adelekan and Jerome 2006). A study of energy use in an informal settlement in the Cape Peninsula in South Africa showed how households that had legal electricity connections and meters could access 50 kWh/month free basic electricity, which encouraged them to cook with electricity rather than paraffin (Cowan 2008).

Low-income urban households often cook with solid fuels that pose serious health threats to household members from indoor air pollution, especially for those with the longest exposure (see Chapter 4 for details). Among urban populations in many sub-Saharan African nations, wood and charcoal are still the most widely used cooking fuels (see, for instance, Ouedraogo (2006) on Ouagadougou, Boadi and Kuitunen (2005) on Accra, Kyokutamba (2004) on Uganda, and van der Plas and Abdel-Hamid (2005) on N'Djaména).

Firewood and charcoal were the major cooking fuels in a sample of 200 low-income households in Jimma, Ethiopia (Ejigie 2008). A survey in Accra found that charcoal was the main cooking fuel for over 70 percent of low-income women (Boadi and Kuitunen 2005). More affluent respondents also cooked with charcoal, though utilization dropped to 40 percent among middle-income and 10 percent for wealthy respondents. Among these respondents, significant health disparities were uncovered. Nearly 30 percent of poor women had respiratory health problems, as compared to 14 percent of medium-income and 3.5 percent of wealthy respondents who were surveyed (*ibid.*). And almost 30 percent of the children surveyed had had respiratory infections during the previous two weeks, of whom 85 percent lived in poor households. Charcoal is the major cooking fuel across income groups in urban areas in Tanzania, Chad, and Uganda (Mwampamba 2007).

This reliance on charcoal by large sections of the population of major (and often rapidly growing) cities generated concerns regarding its contribution to deforestation, although a detailed study in several African nations in the late 1980s found very little evidence of this (Leach and Mearns 1989) and a more recent review suggests that fuelwood is seldom a primary source of forest removal, although 'in some of the areas where charcoal production is concentrated, this may be the case' (Arnold et al. 2006: 606).

Urban dwellers in India are shifting to cleaner cooking fuels, although the shift between 1983 and 1999 was most evident among higher-income groups. In 2000, less than 40 percent of the bottom two urban deciles cooked with clean fuels. And among the lowest income urban groups, adoption of clean cooking fuels hardly increased from 1983 to 2000 (Viswanathan and Kavi Kumar 2005). LPG and kerosene are highly subsidized in India, but non poor groups are the main beneficiaries and many low-income urban residents continue to cook with dirtier energy sources (Gangopadhyay et al. 2005; Pohekar et al. 2005).

In many nations in Asia and Latin America, poor households have switched to LPG but small quantities of other fuels may be purchased to reduce expenses. Expenditures on cooking fuels were high in two peripheral Buenos Aires settlements, comprising 54 percent of total energy budgets (Bravo et al. 2008). Most of the expenditure on cooking was for LPG.

Cooking with LPG is common in the Philippines, but low-income urban households also buy kerosene or biomass fuels to keep costs down. In a survey of two low-income districts in Metro Manila, LPG was the main cooking fuel in 75 percent of households (APPROTECH 2005). However, as LPG prices increased in 2004, low-income groups also began to cook with kerosene, fuelwood, or charcoal. Although residents intended to reduce expenditures, they still paid higher unit prices because they could only afford to purchase small quantities (APPROTECH 2005).

6.5 High household expenditures on energy

A considerable range of national and city studies and studies of particular settlements show how expenditures on fuels for household use are consistently burdensome for low-income households (but energy poverty may not show up as high expenditures or high proportions of income spent on fuel, as discussed above). Examples of high expenditures on energy are:

- In Guatemala, cooking and lighting took up about 10 percent of household expenditures for the bottom three urban deciles in 2000 (ESMAP and UNDP 2003).
- In Thailand, 'slum' dwellers[7] spent about 16 percent of their monthly income on energy (cooking, electricity, transport) in Bangkok and about 26 percent in Khon Kaen. Households in these slums with incomes below the poverty line spent 29 percent of total household income on energy in Khon Kaen and 18.5 percent in Bangkok – mainly because of the high cost of electricity (Shrestha et al. 2008).
- In Ethiopia, fuel and power took 11 percent of expenditure among urban poor (Kebede et al. 2002).
- In Sana'a, the capital of Yemen, the bottom two deciles spent over 10 percent of their incomes on electricity alone (ESMAP and UNDP 2005).
- In Kibera, one of Nairobi's largest informal settlements, for over 100 households surveyed energy expenditures reached 20–40 percent of monthly incomes (Karekezi et al. 2008).
- In Rio de Janeiro (Brazil), many households in surveys in informal settlements were spending 15–25 percent of their incomes on energy (WEC 2006).

- In Cairo (Egypt), households with incomes at the lower poverty line spent 8–20 percent of their income on electricity (Sabry 2009).

Low-income households who obtain electricity through shared electricity meters can be charged higher rates because of rising block tariffs and as the meter records the combined use of several or many households (examples in Kumasi (Ghana), Mumbai, and an informal settlement in South Africa are given by Devas and Korboe (2000) and Cowan (2008)).

6.6 Space heating

Data on heating expenditures are limited, but it is clear that where space heating is needed, low-income urban dwellers can face high costs to keep warm. For instance, surveys in 1999 found that low-income city-dwellers in Armenia, Moldova, and the Kyrgyz Republic devoted 5–10 percent of their household incomes to heating (Wu et al. 2004). Low-income households may also heat their homes with inefficient, polluting fuels to reduce expenditures. During the winter of 2002, Tbilisi's low-income households who were not on the gas network resorted to using wood for heating and cooking (ESMAP 2007). Wood prices were cheaper than those of other fuels, except natural gas. In the heart of South Africa's coal-mining country, residents of Vosman Township rely on coal for space heating, water heating, ironing, and cooking (Balmer 2007). Even in the United Kingdom, four million households were deemed to live in fuel poverty in 2007 (defined by spending 10 percent or more of income on maintaining an adequate level of warmth) (UKDECC 2010).

In China, coal is a key heating fuel for low-income groups, particularly in cold northern cities where heating may take up as much as 40 percent of households total energy needs (Pachauri and Jiang 2008). Although data are not specifically available on coal use for heating, national surveys indicate that 65 percent of the lowest income urban households utilize coal (Pachauri and Jiang 2008). Coal-using urban residents are exposed to extremely high levels of indoor air pollution (Mestl et al. 2007).

6.7 Lighting and access to electricity

Electricity for lighting has important effects on household well-being, including much reduced levels of fire risk (compared to kerosene lights or candles) and much greater convenience from dependable lighting. Adequate illumination also promotes educational activities and supports home-based enterprises, while also improving safety and comfort. Electricity is easily the most efficient lighting fuel and also essential for many appliances

The high proportion of urban dwellers lacking access to electricity in the least-developed countries and in sub-Saharan Africa was noted already. In Table 6.4, there were 15 nations where more than half of urban households did not have access to electricity. Figure 6.2 shows the association between the percent of the urban population with electricity and a nation's average per capita GDP. Note that the per capita GDP range in Figure 6.2 is less than Figure 6.1 since almost all nations with GDPs per capita of $6,000 or more have 95–100 percent of their urban population with electricity. For nations with per capita GDPs below $3,000, there is a quite consistent picture of rising proportions of urban households with electricity, with some variation. There is more variation between US$3,000 and US$6,000.

What is perhaps surprising in Figure 6.2 is the lack of association between a nation's average per capita GDP and the proportion of the urban population with access to electricity for nations with per capita GDPs of less than $2,000. However, perhaps the sample frame for the urban households interviewed in the surveys from which the data in Table 6.4 and Figure 6.2 are drawn did not include the needed proportion of households living in informal or illegal settlements. For instance, half of Kenya's urban population is said to have access to electricity in 2003 yet a survey in 1998 of informal settlements in Nairobi (which housed half of Nairobi's population) found that only 17.8 percent had electricity (APHRC 2002).

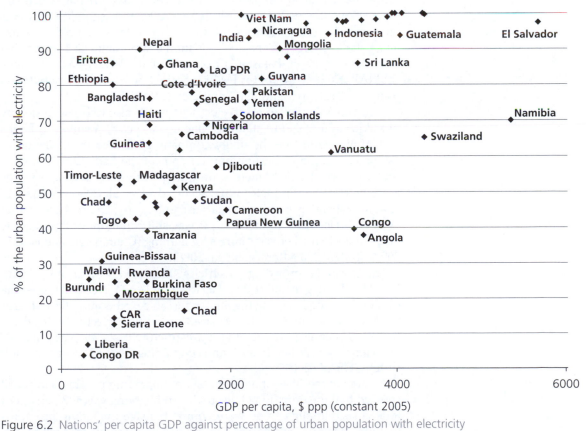

Figure 6.2 Nations' per capita GDP against percentage of urban population with electricity
Source: Data on electricity drawn from UNDP and WHO 2009; data on GDP per capita from World Development Indicators

In most middle-income nations and some low-income nations, most of the urban population has access to electricity. By 2002, there was near-universal access to power in Caracas, Buenos Aires, and Rio de Janeiro (WEC 2006). India's household surveys in 2004–5 found that 91 percent of urban households used electricity; for Chinese city dwellers, household surveys reported that 96 percent used electricity in 2001 (Pachauri and Jiang 2008). Many nations, including Colombia, Dominican Republic, Egypt, Indonesia, Jordan, Pakistan, and Ukraine, report that more than 98 percent of their urban population has electricity (UNDP and WHO 2009). In Table 6.4, Ethiopia, Morocco, Nepal, and Nigeria are among the low-income nations with more than 85 percent of their urban population reported having electricity. These positive developments illustrate that the proximity to existing energy infrastructure in urban areas enables rapid progress, which depends upon implementing dedicated policies of connecting urban poor households. Barriers of low income and limited grid extensions can be overcome. Additionally, a study of energy-use patterns in 'slums' in Bangkok and Khon Kaen in Thailand found almost 100 percent with electricity connections (Shrestha et al. 2008), although in Bangkok 32 percent of households were connected through their neighbors (Shrestha et al. 2008). Almost all 'slum' dwellers in Cairo have electricity connections (Sabry 2009). In Mexico in 2000, access to electricity was enjoyed by 91–97 percent of the lowest-quartile households in cities along the US border (Peña 2005). National surveys in 2001 found that over 80 percent of Pakistan's poorest urban deciles had electricity (ESMAP 2006). Statistics for Delhi report 99 percent of the population using electricity for lighting (Dhingra et al. 2008).

A consideration of access to electricity needs also to consider quality of service in terms of regularity, reliability and cost. A survey in Lagos and Abuja found that the utility company generated their bills using estimated rates, not consumption (Friends of the Earth 2005). In peri-urban Dakar, electricity meters are infrequently read and households are billed on approximate usage levels (Fall et al. 2008). Studies in Harare, Kibera (Nairobi), and urban centers in Yemen showed that even with electricity connections, it was still common for low-income households to use kerosene lamps for lighting (Chambwera and Folmer 2007; ESMAP 2006; Karekezi et al. 2008).

The costs of providing electricity access to low-income urban households are generally quoted as low (Table 6.5). Nonetheless, some caution is needed in using the figures in Table 6.5 because it is not clear whether these are just the cost of extending electricity to these households or also include other infrastructure costs, such as the costs of extending overhead lines and upgrading the power-generation system (USAID 2004).

A study of the costs of different 'slum' upgrading programs in Brazil showed that the provision of electricity and lighting was 1–3 percent of total costs, although these were comprehensive upgrading programs that included provision of water and sewer connections for each house, and building homes for those that had to be rehoused (Abiko et al.

Table 6.5 The cost per household (in current US$) of providing electricity in different cities

City	Cost per household (US$)
Ahmedabad	114
Manila	154
Rio de Janeiro	226
Salvador	350
Cape Town	417

Source: USAID (2004)

2007). The costs would be higher as a proportion of total costs within a more minimalist upgrading program – for instance, one that only provided communal water provision and drainage and not piped water and sewer connections to each household.

Even in cities with well-developed electricity grids, low-income residents may struggle to access power – for instance because the electricity utility refuses to or is not allowed to connect them because they live in an informal settlement that lacks an official address or that lacks needed official documentation. Or connection fees may be unaffordable. In Yaounde, where about 85 percent of households have electricity, 35 percent of electrified households have illegal connections; in some informal settlements, over half the consumers are illegally connected (Tatiétsé et al. 2002). About 90 percent of Dakar (Senegal's capital and largest city) is electrified, but a survey found that a quarter of peri-urban households had illegal connections and in the poorest peripheral areas, between 50–70 percent of electricity connections were illegal (Fall et al. 2008). In part, this was related to high connection fees. Similarly, the upfront cost for electricity connections in Nairobi represented 3–5 months' income for residents of Kibera (Karekezi et al. 2008).

Illegal connections may still imply significant costs to low-income households. If connection is through a neighbor, connection fees and/ or usage fees are paid to them (see Fall et al. 2008 for Dakar, Shrestha et al. 2008 for Bangkok). Studies in Ahmedabad and Manila have shown how households served by illegal connections may pay their landlords more for electricity than if they were legally connected (USAID 2004; ESMAP 2007). In a large informal settlement in the Cape Peninsula in South Africa (Imizamo Yethu), those with informal unmetered electricity connections (typically supplied by an extension cord from a household legally connected) usually paid more per unit consumed (Cowan 2008). Further discussion on energy access issues beyond electricity is contained in Chapter 4.

6.8 Transport

When choosing where to live, low-income individuals or households have to make trade-offs between good locations for access to income-earning opportunities and cost, housing quality, size, and tenure, and provision for infrastructure and services. As discussed above, in most cities in low- and middle-income nations, a significant proportion of low-income groups live in peripheral locations because rents are cheaper and housing is often less crowded or there are more possibilities of obtaining land on which to build housing (although usually illegally). But peripheral locations usually mean high monetary and time costs in traveling to and from work and services. Thus, transportation costs can eclipse household spending on cooking, heating, and lighting.

Various studies of transport use and expenditures in cities or in particular low-income districts show that public transport costs represent a significant part of total household expenditure. For instance, for the inhabitants of eight informal settlements in Cairo, transport costs were a major burden. Many such settlements on the outskirts are not adequately served by the public bus network or the metro. Many inhabitants have to use more expensive privately operated microbuses for part of the journey and a high proportion have to change to other buses or the metro for their journey. High travel costs were one reason why few children went to secondary school (Sabry 2009). Other examples include:

- In Karachi, interviews with 108 transport users who lived in one central and four peripheral neighborhoods found that half were spending 10 percent or more of their income on transport (Urban Resource Centre 2001).
- In Bandung City (Indonesia), interviews with a sample of 145 kampong residents found that nearly 7 percent of their monthly income was devoted to transport costs (Permana et al. 2008).
- In Buenos Aires, a 2002 survey found that the bottom quintile walked to work for 53 percent of their journeys, but they still spent over 30 percent of their family incomes on public transit (Carruthers et al. 2005).
- In Sao Paulo, a 2003 survey found that low-income groups spent 18–30 percent of their incomes on travel (Carruthers et al. 2005). Wealthy residents spent just 7 percent of their incomes on transport, but were able to travel far more frequently. The number of trips completed by Sao Paulo's poor was less than one-third of those completed by the highest-income residents.
- In Salvador (Brazil), low-income residents often live in the urban periphery and a survey of over 500 households in the low-income neighborhoods of Plataforma and Calavera found that transport expenditures averaged 25 percent of monthly expenditures (Winrock International 2005).
- In Khon Kaen (Thailand), a survey of slum households found that on average 9.3 percent of household income was spent on transport (Shrestha et al. 2008).

Thus, it is common in cities for low-income groups to face high transport expenses that curtail their travel possibilities and leave them with onerous journeys, often on foot. Transport costs also limit livelihood opportunities for low-income groups that live in peripheral locations, as the cost and time to reach parts of the city are too high. A 2003 survey in Wuhan, China, showed how prohibitively high transit costs resulted in the poor rejecting jobs far from their homes (Carruthers et al. 2005).

Some studies show how many low-income groups walk long distances to keep their transport expenditures down (see, for instance, Huq et al. (1996) for various cities in Bangladesh, and Barter (1999) for central Bombay/Mumbai and Jakarta). So, while such individuals may pay little for transport costs, they 'pay' through long journey times and extra physical effort. In the survey of Wuhan, China (Carruthers et al. 2005), the bottom quintile reported walking for almost half of their journeys, while 27 percent of their travel was by public transit and 22 percent bicycling.

Informal settlements may not be served by public transit, and low-income women can face particular challenges in accessing secure, efficient transportation (Watkiss et al. 2000). A large survey in Nairobi found that 67 percent of low-income women living in 'slums' walked, significantly higher than the 53 percent of their male counterparts (Salon and Gulyani 2010). Informal buses have proliferated in many cities, and can help alleviate transport shortages (Zhou 2000). However, in this unregulated sector vehicles are usually old and overcrowded, accidents are common, and customers are vulnerable to rising or erratic fares.

6.9 Differentials within urban populations

The reliance of national governments and international agencies on sample surveys for much of their social data produce aggregate statistics that conceal the inaccessibility of clean, affordable energy for low-income households in urban areas. Urban areas generally have higher proportions of their populations served with LPG and electricity than rural areas, but most low-income households within urban areas in low- and middle-income nations struggle to afford these modern fuels. India's national surveys found that in 2000, 7–15 percent of the bottom two urban deciles used LPG versus nearly 70 percent in the top two deciles (Gangopadhyay et al. 2005). Access to electricity, for instance, may be limited to low-income residents because it is only available to those with formal addresses (and thus exclude 30–60 percent of the population in many cities). Utility companies often refuse to extend electricity to informal settlements because of the uncertainty of who is the 'account holder' or because they are not allowed to do so.

Some studies show how it is common for low-income urban households in low- and middle-income nations with electricity connections to use 20–60 kWh/month (see Kulkarni et al. 1994; Karekezi et al. 2008). This is a small fraction of average household use in the United States (640–1329 kWh/month depending on the region) or

Europe (341 kWh/month). So it is likely that differentials of the order of 100 or more are present between the world's wealthiest and least wealthy households with electricity. But there are also large differentials within cities in low- and middle-income nations; for instance, Kulkarni et al. (1994) recorded electricity use among low-income households in Pune at 33 kwh/month and electricity use among high income groups of 293 kwh/month. Pachauri et al. (2004) considered how the amount of energy consumed and the quality of energy services available varied by income group (Table 6.6).

6.10 Summary

As noted in the introduction, the existing development literature on basic needs has focused principally on provision of water and sanitation, health care, education, and sufficient food. It rarely considers clean energy and electricity – as can be seen in the lack of attention to this in the 8 Millennium Development Goals. However, several hundred million urban dwellers in low- and middle-income nations lack access to electricity and are unable to afford cleaner, safer fuels such as gas or LPG (or even kerosene). Most are in low-income nations in Asia and sub-Saharan Africa. In many such nations, more than half the urban population still rely on polluting cooking fuels (resulting in large health burdens) and necessitating additional time to obtain fuels. In many low-income nations, more than half the urban population also lacks access to electricity, even though urban population concentrations can potentially decrease the unit costs for providing electricity. Urban concentrations also lower the cost of providing gas (or LPG gas distribution). But at present, a high proportion of urban dwellers in low- and middle-income nations find it difficult to afford their 'energy bills' (for fuel and, where available, electricity and expenditure on

Table 6.6 Grouping households in India by the amount of energy they consume and the energy services available to them (average household of five persons) in Watt-years (1 Wyr =31.55 MJ)

Energy services of households	Useful energy consumption per capita (Wyr)
Associated with extreme poverty: up to one warm meal a day, a kerosene lamp, possibly a little hot water	<15 W
Associated with poverty: 1–2 warm meals per day, a few kerosene lamps or one electric bulb, some hot water	15–30
Associated with above the poverty line: two warm meals a day, hot water and lighting, some small electrical appliances for groups with electricity, possibly a scooter	30–60
Associated with a comfortable level of well-being: two or more warm meals a day, hot water, lighting, some space heating, for groups with electricity possibly some space cooling and electric appliances, possibly a scooter or an automobile	60+

Source: Pachauri et al. (2004)

transport); these often take 15–20 percent of household income and for many a higher proportion.

However, in many middle-income nations (and all high-income nations) nearly all low-income urban dwellers have legal electricity connections and can afford clean fuels. The shift to clean fuels and the availability of electricity bring many advantages in terms of health, convenience, and time saved in accessing and using energy – for no time is needed to purchase or gather fuels, and then carry home solid or liquid fuels or LPG cylinders. As noted in the introduction, reliable electricity supplies also bring many other obvious advantages – reliable, cheap, and safe lighting at night, the possibility of fridges, televisions, and electric fans, support for home enterprises, and a very large reduction in fire risk.

The costs of connection to an electricity grid and the use of electricity can be burdensome for low-income groups, but innovations have reduced these costs – for instance, rising tariffs with low prices for 'lifeline' consumption (or in South Africa no charge for up to 50 kWh/month), pay-as-you-use meters, and standard 'boards' that have one or two plug sockets and a light and that remove the need for household wiring. South Africa pioneered ready-made boards beginning in 1990 and prepaid meters (Bekker et al. 2008). Under Malawi's Mbayani electrification program, Blantyre residents pay for their compact boards through a monthly tariff over five years (ESMAP 2007).

The urgent need to reduce greenhouse gas emissions globally might be considered a constraint on low-income urban dwellers obtaining electricity and clean fuels. However, the shift from dirty fuels to clean fuels produces a lower than expected contribution to global warming because of the inefficiencies in how dirty fuels are consumed and in the reduced contribution of fuel use to black soot aerosols. Thus, a shift to clean fossil fuels leads to major improvements in the global impacts associated with non-CO_2 emissions. In addition, current differentials in electricity use per household or in CO_2 emissions per household are likely to vary by a factor of at least 100 between the wealthiest households and the least-wealthy households (Satterthwaite 2009).

The constraints on supporting the shift to clean fuels and providing all urban households with electricity are less in energy policy and far more in government policy and daily practice. A large part of the urban population that lacks clean energy and electricity also live in informal settlements with insecure tenure and a lack of reliable piped water supplies and good provision for sanitation and drainage. They often lack access to schools, health care services, and emergency services. Governments often ignore them, even though these settlements frequently house 30–60 percent of cities' populations, most of its low-wage labor force, and many of its enterprises. It is mostly in nations where relationships between local government and the inhabitants of these informal settlements are not antagonistic, with widespread public support for 'slum' and squatter upgrading, that clean energy and electricity successfully reaches low-income urban groups.

Notes

1   This includes a shift to 'clean' fuels – clean in the sense of minimizing pollution and health impacts for the users – for instance, with electricity and gas or energy derived from renewable energy sources being 'clean' and coal and raw biomass being 'dirty' (how dirty these are depends on the technology used in the home). Kerosene and charcoal fall between these two extremes. The term 'clean fuels' is ambiguous in that it is used to mean different things – for instance, for fuels or energy sources that have low or no CO_2 emissions, rather than lower health impacts for users. In addition, electricity at the point of use may be 'clean', but it often comes from coal-fired power stations that have high CO_2 emissions and often high levels of pollution.

2   This chapter does not cover issues of energy poverty in high-income nations and in low- and middle-income nations that were formerly part of COMECON (termed countries undergoing economic reform in the Global Energy Assessment).

3   However, there are also serious limitations in the detail and accuracy of statistics on provision for water and sanitation within urban areas – see Satterthwaite 2010.

4   UNDP and WHO (2009) state that modern fuels include electricity, liquid fuels, or gaseous fuels and so include liquid petroleum gas (LPG), natural gas, kerosene (including paraffin), ethanol, and biofuels, but exclude all traditional biomass (e.g., firewood, charcoal, dung, and crop residues) and coal (including coal dust and lignite).

5   This source is inconsistent in how it reports some of the figures for access to electricity; the figures above for the least-developed countries and sub-Saharan Africa are from Figure 3, but the accompanying text (page 12) says that 46 percent of the urban population of least-developed countries and 56 percent of urban dwellers in sub-Saharan Africa lack electricity access. The report does not specify where its population figures come from, although it lists the UN Population Division's World Population Prospects: the 2006 revision in its sources.

6   This issue is discussed in more detail in the section on lighting and electricity below (6.7).

7   We try to avoid using the term 'slum' because of its derogatory connotations. Classify a settlement as a slum, and it helps legitimate the eviction of its inhabitants. In addition, the term slum is often used as a general term for a range of different kinds of housing or settlements, many of which provide valuable accommodation for low-income groups and that do not need replacement but provision for infrastructure and services. But it is difficult to avoid the term slum for at least three reasons. The first is that it is used in studies from which we want to draw. Second, some urban poor groups have organized themselves as slum dweller organizations or federations. In some Asian nations, there are advantages for residents of informal settlements in being recognized officially as a 'slum'; indeed, the residents of such settlements may even lobby to become a 'notified slum'. The third is that the only global estimates for deficiencies in housing collected by the United Nations are for 'slums'.

7

Energy demand and air pollution densities, including heat island effects

Niels B. Schulz, Arnulf Grubler, and **Toshiaki Ichinose**

7.1 Introduction

The concepts of energy demand and pollution densities refer to the *amount of energy* used and/or produced or the pollution emitted *per unit of land*. Their common denominator and driver is urban population density. Despite being of fundamental importance in an urban context, the literature on energy or pollution densities is surprisingly limited, apart from that on urban heat island effects, reviewed in detail in section 7.4 below.

7.2 Urban energy demand densities

This section illustrates the concept of energy supply and demand densities both generally and by drawing from contrasting examples of two high-density megacities (Tokyo and London), as well as a small, low-density city (Osnabrück, Germany). A brief discussion of associated policy issues follows.

The classic reference on energy demand and supply densities remains Smil (1991), from which Figure 7.1 is adopted in modified form.

The customary unit for energy densities is watts per square meter (W/m^2), referring to an energy use per year equivalent to the power of one watt.

Within an urban context particularly, energy demand densities are of significance. The twin influences of high population and high income mean that the spatial density of energy demand of cities typically ranges from 10–100 W/m^2, a range exemplified by cities such as Curitiba (Brazil) and Tokyo. Energy-demand densities in smaller portions of urban areas can approach values of 1000 W/m^2, as in sub-sections of the twenty-three wards of central Tokyo (Dhakal et al. 2003).

The significance of urban energy-demand density arises in three areas. First, the higher the energy density, the larger the impact of emissions, either as air pollutants or as waste-heat releases. Second, from an energy-demand perspective, high energy densities suggest opportunities for waste-heat recycling and economic provision of clean district heating and cooling services. Third, energy-demand densities

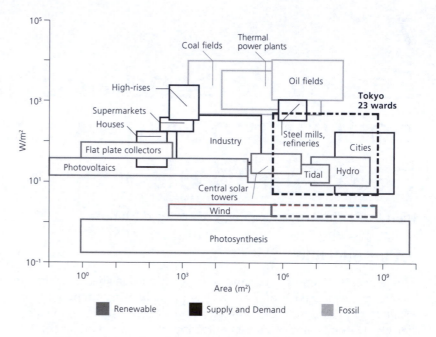

Figure 7.1 Energy densities of energy supply from fossil and renewable sources (gray) versus density of energy demand (black) for typical settings, in W/m² and m² area

Source: modified from Smil (1991)

are significant constraints for the provision of energy services through renewable energies, which (with the exception of geothermal) typically range from 0.1 to 1 W/m² and thus yield a significant mismatch between demand and supply at the urban scale.

From an energy-systems perspective it is important that the prevailing high energy-demand densities characteristic of urban areas are much in line with those of fossil fuel infrastructures and conversion devices (Grubler 2004). The general mismatch between (high) urban energy demand and (low) renewable energy supply densities is shown with actual energy demand data for London and Tokyo in Figure 7.2.

The typical order of magnitude of energy use of a megacity is in the order of an exa-Joule (10^{18} J), a unit normally reserved for reporting the energy use of entire countries. The (direct final) energy use of Tokyo's twenty-three wards is estimated to be about 0.6 EJ and that of the larger Tokyo Metropolitan area as 0.8 EJ (Tokyo Metropolitan Government 2006), compared to 0.6 EJ for London (thirty-three boroughs) and 0.8 EJ for New York City[1] (Kennedy et al. 2009). Energy-demand densities in Tokyo and London typically span a range from a few W/m² to >200 W/m², as in the City of London or in the top twenty-five grid cells (i.e., top 25 km²) of the Tokyo wards that use close to 18 percent of Tokyo's total final energy. Such high energy-demand densities are comparable to the entirety of the solar influx, which equals 157 W/m² in Tokyo and 109 W/m² in London. Mean energy densities, 28.5 W/m² for the Tokyo 23 wards (621 km²) and 27.4 W/m² for Inner London (319 km²), are similar. (For Greater London with its larger size (1,572 km²), lower population densities, and greater extent of green areas, energy densities are naturally lower, at 13 W/m².)

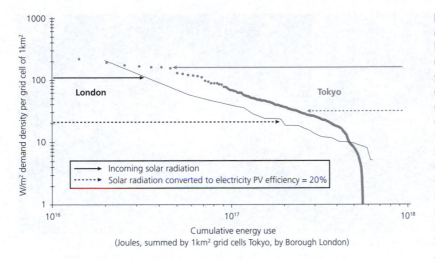

Figure 7.2 Energy-demand densities (W/m²) for London (33 boroughs, black line) and Tokyo (1 km² grid cells, 23 wards, gray symbols) versus cumulative energy use of these spatial entities (in joules). For comparison, the energy flux of incoming solar radiation (W/m², solid line arrows) and the electricity that could be generated (assuming photovoltaics (PVs) with a conversion efficiency of 20%, dashed line arrows) is also shown

Sources: Dhakal et al. (2003); UKDECC (2010)

Assuming that all the incoming solar radiation could be converted for human energy use (e.g., to electricity with 20 percent efficient PV panels), the maximum renewable energy supply density would range from 22 (London) to 31 (Tokyo) W/m² in line with average demand densities in the two cities, but only under the assumption that the entire city area could be covered by PV panels along with interseasonal storage! Even assuming an upper bound of potential PV area availability (roofs, etc.), assuming numbers characteristic for a low-density urban area (Osnabrück, Germany, see below) of 2 percent of the city area, solar energy could provide a maximum of between 0.4 (London) and 0.6 W/m² (Tokyo), which would cover between 2 percent (Tokyo's twenty-three wards) and 1–3 percent (Inner to Greater London) of urban energy use in the two cities. *Local renewables* can therefore only supply urban energy in niche markets (e.g., low-density residential housing), but can provide *less than 1 percent* only of a megacity's energy needs.[2]

Given that local renewables in large cities are at best marginal niche options (because of the density mismatch between energy demand and supply), what is their potential in small, low-density cities? Using aerial survey techniques, Ludwig et al. (2008) performed a comprehensive assessment of suitable application of rooftop solar PVs for Osnabrück (Figure 7.3). Osnabrück, with an area of 120 km² and a population of 272,000 (a density of 23 people/hectare) is characterized by an incoming solar radiation of 983 kWh/m² (112 W/m²). In the study, all suitable roof areas of some 70,000 buildings in the city were assessed (considering optimal inclination as well as shadowing by adjacent buildings) and the results published for local residents in a database per individual dwelling.

The study identified a total of 2 million m² of suitable PV roof area for Osnabrück (corresponding to 1.6 percent of the city area), which if used completely for PV applications could provide some 249 million KWh of electricity, or about the entire *residential* electricity demand of the city (235 million kWh) or up to 26 percent of the total electricity demand of Osnabrück (940 million kWh). It is of particular interest to interpret the

Figure 7.3 Example of assessing local renewable potentials: suitable roof-area identification for solar PV applications for the city of Osnabrück, Germany. Red: roof area well suited for PV; orange: suitable; yellow, only conditional suitability for PV applications; gray: shadowed roof area (unsuitable). (See color plate 8)
Source: modified from Ludwig et al. (2008)

Osnabrück results in terms of their corresponding renewable energy supply density, which adds up to some 0.2 W/m^2 and can be considered a realistic upper bound of the local renewable energy potential for low-density urban areas. In the example of Osnabrück, local renewables could provide some 3.3 GJ/capita or 2 percent of the average German per capita final energy use of 154 GJ/capita.

This example shows an important trade-off between population density, transport energy demand, and the potential for local renewables. Generally, the areas available for harvesting local renewable energy flows are higher for a *lower* population density urban area. Osnabrück, with a population density of 23 inhabitants/ha and a high proportion of low-density residential housing (single-family homes) evidently offers larger potentials to harness solar energy compared to a megacity with population densities of 130 people/ha and predominantly high-rise buildings (as for Tokyo's twenty-three wards). However, this higher potential for harnessing local renewables at lower population densities is at odds with the potential to lower the dependence on energy-intensive individual transport modes (automobile usage) in urban areas via public transport. Public transport systems require relatively high population densities to offer an attractive and economically viable alternative to private automobiles, with the minimum population density threshold required typically above 50–100 inhabitants/ha. In terms of energy, there is thus an inherent trade-off between urban form, transport choices, and the potential of harnessing local renewable energy flows. Put simply, the positive energy implications of an 'active' building (e.g., a 'Passivhaus' standard energy-efficient home with PV panels on the roof that produce electricity both for its own use and for the grid) can turn negative if the building is situated in a low-density, suburban setting with a high automobile dependence.

Therefore, if renewable energies are increasingly to supply the urban energy needs on a large scale, the resulting needs for conversion and long-distance transport, as well as very large energy 'catchment' areas (the 'energy footprint' of cities), needs to be taken into account.

In an attempt to quantify the implications of the energy supply and demand–density mismatch, IIASA researchers used spatially explicit energy-demand estimates for Europe to calculate related energy-demand density zones (Figure 7.4). The study found that about 21 percent of final energy demand in Western Europe is below the supply density threshold of 1 W/m^2, a characteristic upper bound for locally harvested renewable energy flows. The corresponding value for Eastern Europe is somewhat higher, with 34 percent of energy demand below 1 W/m^2. Nonetheless, in all densely populated, highly urbanized regions, the majority of renewable energy supply has to come from areas of low population and energy-demand densities, where renewable energy flows can be harnessed and transported to the urban energy-use centers, which represents a formidable infrastructure challenge.

The findings of the IIASA study are also confirmed by a detailed, spatially explicit assessment of solar electricity (PV) potentials for all of Western Europe by Scholz (2010, 2011).

The Scholz study identified a total solar (rooftop)[3] PV generation potential of 638 TWh (equivalent to 2.3 EJ, or some 40 percent of the residential electricity demand in Western Europe, and 23 percent of the total electricity demand in the region). Of that solar PV potential, 637 TWh (99.8 percent) is below a maximum energy supply density of 0.5 W/m^2, and 563 TWh (88.2 percent) below a energy supply density level of 0.2 W/m^2 (Scholz 2011). Renewable energy supply densities in urban areas are therefore maximum in the range of 0.2 to 0.5 W/m^2 which are thus between 2 to 5 percent of characteristic urban energy demand densities of 10 W/m^2.

7.3 Pollution densities

A corollary of energy densities is that of pollution density. High population density also leads to high *exposure*[4] density to pollution risks.

However, at least for traditional air pollutants such as particulates, urban pollution exposures also need to be seen in context, as only approximately one-third of the global pollution exposure is urban, whereas two-thirds are rural, because of the dominance in global particulate pollution exposure of indoor air pollution in rural households of developing countries (Table 7.1). Smith (1993) developed the concept of global exposure equivalent (GEE), which represents a renormalized index of the global summation of pollution exposure (pollution concentration times population exposed) calculated for a range of human environments. According to Smith (1993), global human exposure to traditional pollutants is dominated by indoor air pollution in rural and urban households in developing countries as a result of the continued use of traditional biomass for cooking.

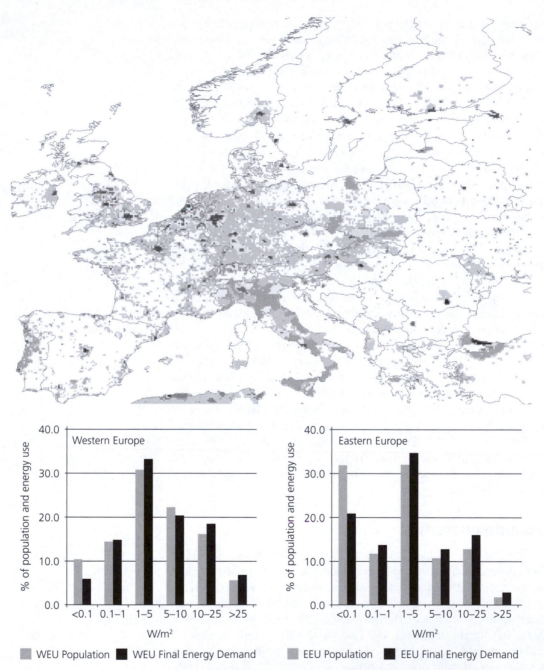

Figure 7.4 (**Top**): Spatially explicit energy demand densities in Europe (W/m²): blue and white areas indicate where local renewables can satisfy local low-density energy demand (<0.5–1 W/m²); yellow, red, and brown colors denote energy demand densities above 1, 5, 10, and 25 W/m² respectively. (**Bottom**): Distribution of population (gray) and final energy demand (black) (in percent) as a function of energy demand density classes in W/m² for Western Europe (left panel) and Eastern Europe (right panel). Only 21% (Western Europe) and 34% (Eastern Europe) of energy demand is below an energy demand density of 1 W/m² amenable to full provision by locally available renewable energy flows. The high energy densities of cities require vast energy 'hinterlands' that can be 100–200 times larger than the territorial footprint of cities proper requiring long-distance transport of renewable energies. (See color plate 9)

Table 7.1 Global exposure equivalents to particulate emissions. Note, in particular, the continued dominance in developing countries of indoor air pollution from traditional biomass cook stoves compared to the urban outdoor air pollution exposure

	Concentrations (µg/m³)		Exposures (GEE)[a]		
Group of Nations	**Indoor**	**Outdoor**	**Indoor**	**Outdoor**	**Total**
Developed					
Urban	100	70	5	< 1	**6**
Rural	60	40	1	< 1	**1**
Developing					
Urban	255	278	19	7	**26**
Rural	551	93	62	5	**67**
Total			**87**	**13**	**100**

Source: adapted from Smith (1993)

For more modern forms of pollution (sulfur and nitrogen oxides (SO_x and NO_x) and ozone (O_3)), the corresponding GEEs have not been calculated, but it is highly likely that the respective role of indoor versus outdoor air pollution as the main source of a population's pollution exposure risk is reversed; that is outdoor air pollution and urban settings comprise the dominant form of pollution exposure. As an example, consider emissions of sulfur dioxide (SO_2). The 'hotspot' of sulfur emissions and pollution, which has for decades been the 'black triangle' (the coal-rich border area of Poland, the Czech Republic, and East Germany) in Europe, was remediated by successful European sulfur-emission reduction policies. The current sulfur-emission hotspot is now in China (Figure 7.5), where high elevated levels of sulfur emissions particularly affect the urban populations and triggered policy responses (see also Chapter 12).

From an environmental perspective high urban energy demand and the resulting pollution densities hold two important implications. First,

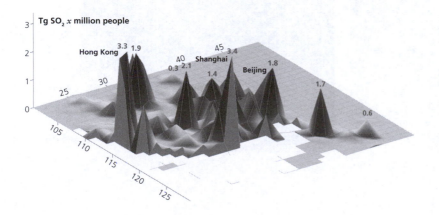

Figure 7.5 Human exposure to sulfur emissions (population in million x emissions x Tg SO_2, z-axis) in China (2000), based on an analysis of gridded socioeconomic and emission data. (Units on x, and y-axis refer to geographical longitude and latitude). Note the high pollution exposure in major urban areas of China
Source: IPCC RCP scenario database (IIASA 2010)

energy use usually involves heat losses at well above ambient temperatures and high densities of urban energy use also imply high densities of urban waste-heat releases. These combined with the (high) thermal mass of buildings in densely built-up urban land give rise to the 'urban heat island effect' (see below) in which urban mean temperatures are several degrees higher than those of surrounding hinterlands.

Second, fuel choice becomes of paramount importance: pollution-intensive fuels (biomass or coal) used at the high demand densities of urban areas quickly result in unacceptably high levels of pollution concentration (such as the London 'killer smog' of 1952 or the current air-quality situation in many cities, especially in the developing world). Even low-pollution fuels, such as natural gas, can quickly overwhelm the pollution dissipative capacity of urban environments. So, high energy-demand density requires zero-emission (at the point of final use) fuels: electricity and perhaps, in the long run, hydrogen.

7.4 Urban heat island effects

7.4.1 Formation of urban heat islands

Urban heat islands describe the frequently observed pattern of urban air temperatures that exceed those of neighboring, more rural areas. In temperature maps, which delineate neighborhoods of similar temperature with contour lines ('isotemperatures' 'isothermals'), urban areas stand out as 'islands' that form 'heat domes.' For example, Figure 7.6 shows these for Tokyo, the city for which most literature on the heat island effect is available.

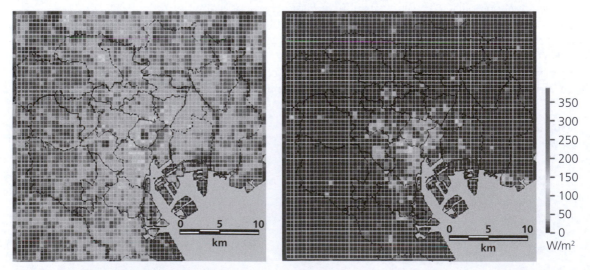

Figure 7.6 Sensible (left) and latent (right) anthropogenic heat[5] emission in Tokyo (W/m²). (See color plate 10)
Source: Ichinose (2008)

Urban temperatures typically peak some hours after midday, but the absolute temperature difference against rural areas can be even larger during the night under cloud-free conditions. Heat islands are facilitated in climatic situations of low air movement. Wind otherwise disperses temperature plumes. Heat islands are similar to air-pollution concentrations and they can be enhanced by local topography and climate patterns that prevent mixing of the boundary layer. In terms of average temperature difference, urban areas are often 1–3°C warmer than the surrounding air, and at individual locations in calm and clear nights temperature differences can exceed 12°C (Klysik and Fortuniak 1999). With increasing energy use, the extent of urban heat island effects increases (Figure 7.7), which results in local warming.

Heat islands are, among other factors, caused by urban energy use through anthropogenic heat release. Without planning or intervention strategies there is a risk of maladaptation feedbacks, in which heat island countermeasures trigger increasing energy use, which amplify the heat island (Figure 7.8). The increasing use of air-conditioning equipment in buildings and automobiles (which dump waste heat into the atmosphere) are one such example (Kikegawa et al. 2003; Crutzen 2004).

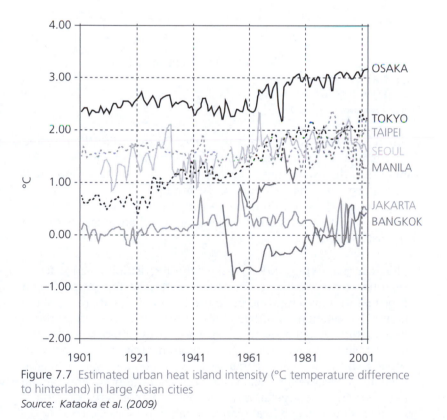

Figure 7.7 Estimated urban heat island intensity (°C temperature difference to hinterland) in large Asian cities
Source: Kataoka et al. (2009)

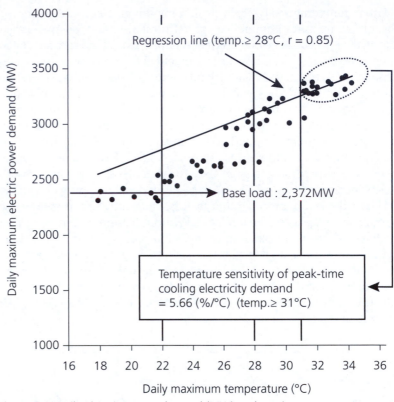

Figure 7.8 Daily electric power demand (MW) and maximum temperature (°C) in Tokyo, June–August 1998
Source: Kikegawa et al. (2003)

Urban heat islands also affect the energy system directly. Elevated environmental temperatures reduce the efficiency of thermal power plants. Also, the availability of cooling water for thermal or industrial plants can be reduced as water bodies warm up.

A range of factors contribute to the formation of urban heat islands and their relative contribution varies among urban areas (Seto and Shepherd 2009):

1. The geographic context defines the natural radiation balance across seasons, temperature, precipitation patterns, topography, and exposure to prevailing wind patterns. In humid climate regions, the natural vegetation is often forest dominated. Urban surface-temperature variation is then, at least partly, moderated by the latent-heat transfer through evapotranspiration of adjacent vegetation (Oke 1987). Parks with water bodies and extensive vegetation cover reduce heat island formation. In more arid climate regions with low vegetation cover, daily temperature variation is more pronounced and heat island formation is more likely.
2. The replacement of natural vegetation with artificial surface materials for buildings, squares, and transport infrastructure results in more

incoming radiation being stored during daytime, particularly if materials are dark, such as bitumen and asphalt. The albedo changes and differences in specific heat capacity of construction material result in more incoming energy being accumulated in surface material during daytime, which is later emitted as infrared radiation at night (Taha 1997). Thermal insulation of buildings can reduce their specific heat capacity drastically.

3. Urbanization modifies the local hydrology. Natural groundwater recharge is typically prevented by the extensive use of impermeable surface materials. Rain gutters, drainage systems, sewers, and canals channel precipitation rapidly away to avert urban flooding. Additionally, the extraction of water from local wells and for construction projects often lowers water tables, which results in lower water availability for the remaining vegetation, and consequently lower evapotranspiration and associated cooling through latent heat transfer.

4. Also, structural characteristics of the urban form (Weng et al. 2004) affect the efficiency of heat loss through radiation and convection. Narrow street canyons with a limited sky view prevent heat loss through direct radiation upward. The urban layout and orientation of street corridors in relation to prevailing wind patterns affect the efficiency of heat transfer through convection and boundary layer mixing. These factors can be addressed through planning regulations. In Germany, for example, the concept of urban ventilation pathways (Luftleitbahnen) aims to maintain radial corridors of cold winds to reach urban centers.

5. The metropolitan area size (extent) amplifies the magnitude of the urban heat island effect. For some rapidly growing urban areas, such as Los Angeles or Kobe, a continuous increase in urban heat island temperature of up to 0.5°C/decade occurred over the past sixty years. This trend is partly amplified by changes in energy-use patterns (Böhm 1998) and needs to be compared against general trends of global surface temperature warming of about 0.7°C over the past century (Oke 2006; IPCC 2007; Kataoka et al. 2009) (see Figure 7.7).

6. As energy demand is concentrated particularly in urban centers, the consequential release of anthropogenic heat is similarly dense in these areas. Industrial and service-sector activity, residential housing, and transport functions are typically clustered in close proximity. Electricity use and combustion processes in buildings and vehicles, for heating and cooling, lighting, or motion, all result in vast quantities of waste heat being released (Rosenfeld et al. 1995). To a small extent, the metabolic activity of biological body functions of the human population also contributes to this.[6] Global average estimates attribute a resulting climate forcing of about 0.028 W/m² to anthropogenic heat release (technical and biogenic). In North America and Europe these figures are estimated to be higher, at +0.39 and +0.68 W/m², respectively (Flanner 2009). For the Ruhr area in Germany, average anthropogenic heat-related forcing values of 20 W/m² were calculated by Block et al. (2004). At higher spatial and

temporal resolution, the values are much larger, often between 20 and 100 W/m². Numerical simulations for heat discharge of individual neighborhoods in the Tokyo metropolitan region, for example, indicate radiative forcing values of up to 700 W/m² during the day and in summer time (Dhakal et al. 2003). However, urban heat islands do not always increase urban energy demand. In higher latitudes the resulting reduction in heating demand in winter can more than compensate the additional cooling energy demand in summer. Integrated climate-energy system models increasingly aim to capture such effects (Kanda 2006; Oleson et al. 2010).

7.4.2 Mitigation

Strategies for heat-island mitigation include behavioral and technological solutions. They can provide various co-benefits, including energy savings, peak-load reduction, air-quality improvements, and beneficial health, psychological, and socioeconomic effects.

Building design and layout allow solar gains of houses in summer to be minimized and increased passive gains during winter (e.g., in Passivhaus designs). Reduction in cooling demand can also be achieved through the use of deciduous vegetation for shading (including vertical greening of facades) or the application of mechanical shades, shutters, or 'smart glass windows,' with modified transmission properties of heat and light on demand. Albedo changes and the use of reflective paint and surface material on roofs and transport infrastructure is another particularly cost-effective mechanism to reduce heat absorption in urban areas. Improving the insulation of the building stock to prevent the warming and storage of solar influx in material of high specific heat capacity, such as concrete, is another measure, just like the expansion of shading structures (vegetation or textile) in general. The shading of parking lots, for example, not only provides thermal comfort, but also reduces emission of volatile organic compounds (significant precursors of O_3 formation) from parked vehicles.

Active cooling via enhanced evapotranspiration can be induced through ponds or fountains, green roofs, and tree planting, or through the generation of artificial mist (microdroplets of water) to create local cooling clouds. In preparation for urban heat waves, the city of London considered the need to prepare 'cooling shelters' for vulnerable or elderly population, as the typical UK housing stock is not equipped with air-conditioning (City of London 2007). This is a method already used in some parts of the world where extreme high temperatures are sometimes experienced.

Changing the timing of activity patterns to avoid the hottest hours of the day was a traditional response to hot climates. In peak-load management programs some of this rationale is revived. Operators of office space in New York City are given price incentives to start air-conditioning in the early hours of the day to reduce demand during peak hours of electricity demand (Bloomberg 2007). While the primary motivation was to reduce peak power demand in a grid of constrained capacity, this measure also reduces the peak in waste-heat emission in the early afternoon hours. While probably not the most suitable for dense, high-rise developments such as Manhattan, solar cooling devices that provide a maximum cooling output at periods of peaking outdoor temperatures appear to be suitable for low-density cities.

In 2005, the Japanese environmental ministry started a campaign titled 'Cool biz' to restrain the use of air-conditioning units to an indoor temperature of 28°C. The campaign aimed to loosen the strict dress code of full-suit, tie, and long-sleeved shirts for office workers during the hot season (June 1 to September 30) and promoted a more heat-tolerant dress code of short-sleeved shirts without ties (Pedersen 2007).

Notes

1   Final energy use within the city limits and excluding bunker fuels (aviation, shipping). The latter are reported to be 0.28 EJ (0.2 EJ aviation fuel and 0.08 EJ marine bunkers) for New York City compared to 0.76 EJ final energy use in 2005 (Kennedy et al. 2010). For London, aviation fuel also accounted for some 0.2 EJ for the year 2000 (Mayor of London 2004).

2   This mind experiment considers a highly efficient conversion route of solar energy via high-efficiency PVs (with 20 percent net conversion efficiency). Assuming biomass as an alternative reduces the energy yield by a factor of up to 20, as the average conversion efficiency of solar energy via photosynthesis is only around 1 percent. Conversely, considering solar hot-water collectors (with a maximum efficiency approaching 100 percent of incoming solar energy) also does not change drastically the conclusion of the extremely limited local renewable potentials in high density cities, as solar hot water typically provides only a few percent of energy demand (hot water accounts for 2 percent of final energy demand in Europe (Eurostat 1988)). Even if this were provided entirely by solar energy where feasible (in low- to medium-density housing, as high-rise buildings do not offer sufficiently large roof areas) the yield is less than 1 percent of energy demand in a densely populated large city.

3   Adding also building facades to the potential PV areas does not change the results significantly. In a study of solar PV potentials considering the entire building envelope Gutschner et al. (2001) estimated a total electricity potential of 600 TWh for a sample of ten European countries, which is good agreement to the Scholz study (638 TWh). Facades were estimated to add another 25% to the rooftop PV potentials estimated by Gutschner et al. (2001).

4   Exposure risk: product of population × pollution level × exposure time of population to pollution.

5   Sensible heat flux: air is heated directly by the heated ground surface. Latent heat flux: evaporation from wet ground surface or from cooling towers settled on top of buildings and evapotranspiration from vegetation. This type of energy exchange does not change air temperature. Its energy is consumed in the phase change from water to moisture.

6   Assuming about 100 W of biological energy use per person and maximum population densities of 40,000/km² in some cities in developing countries, this factor can contribute up to 4 W/m² additional forcing. Typical urban population densities are lower.

8

Supply constraints and urban policy frameworks

David Fisk

8.1 Supply constraints, including reliability and security

Cities depend on extended energy networks and failures can occur on a regional and national scale. On August 14, 2003, a cascading outage of transmission and generation facilities in the North American Eastern Interconnection resulted in a blackout of most of New York State as well as parts of Pennsylvania, Ohio, Michigan, and Ontario, Canada. On September 23, 2003, nearly four million customers lost power in eastern Denmark and southern Sweden following a cascading outage that struck Scandinavia. Days later, a cascading outage between Italy and the rest of central Europe left most of Italy in darkness on September 28. These major blackouts are among the worst power-system failures in the past few decades. They had a profound effect on power-system philosophy because these networks were some of the world's most sophisticated power-generation distribution systems. In particular, the US failure was promoted by an early underlying failure in software used to control networks, which meant the scale of the emerging problem was recognized too late to prevent the cascading failure.

Energy efficiency and energy resilience are not entirely coincidental outcomes. Thus, a low-energy settlement might be even more sensitive to disruptions in supply than a settlement with some slack in its energy system because the low energy settlement is always operating at its minimum. A dense high efficiency city with no power for its elevators may be in a worse state than a low-density urban settlement without power. Renewable power sources based on wind or solar alter the reliability profile. As they are of a much smaller unit size, they do not induce major dropouts, as happens when a large nuclear power station needs to come offline very rapidly. While often termed 'intermittent' this term is better reserved for conventional plant failure. A better term would be 'variable'. Conversely, the variability in available power they supply may require them to be shadowed by a rapid-response plant. They may also exacerbate failure cascades because of switching out for self-protection when the power system is stressed heavily. Urban energy looks destined to be less reliable than it once was in the developed world of the 1960s.

The increasing dependency on power even for simple clerical tasks, let alone for critical functions such as hospitals, means that stand-by power supplies could be an increasing feature in urban systems. One radical suggestion (Patterson 2009) is that it is possible for the local distributed power generation to become dominant and the national distribution systems only handle back-up. This is already effectively the case for dwellings that use micro-generators for power and heat. Another suggestion is that more sophisticated metering and tariffs could incentivize the extension of demand-side load management from large facilities of 'interruptible supply' at the micro scale. In line with the efficiency–resilience argument it is expected that vulnerability to societal interruption is higher in countries with generally very secure supplies in which the economy has sought an equilibrium that assumes secure power supplies than in countries with frequent brownouts and blackouts in which the economy has adjusted to coping with the risk.

The winter of 2008/2009 in Europe showed the vulnerability of gas-supply networks to urban areas. Gas can be stored under pressure both 'packed' in the mains and in dedicated storage facilities, but gas is currently supplied directly to consumers and power generation, and so indirectly to electric mass-transit systems. Thus, a failure of supply pressure has wide implications. Coastal cities can increase their robustness with liquid natural gas (LNG) terminals, but LNG is a world-traded product and may come at a high price in a regional emergency.

Liberalization of gas and electricity market pressures gives the lowest prices to consumers, but the effect is to incentivize producers to 'sweat' their existing assets (e.g., Drukker 2000). It may then become necessary to introduce further complexity into tariffs to incentivize investment and reflect the value to the consumer of lost load. Undercapitalization of energy networks, many of which were built in the 1960s, remains a real risk over the next twenty years in developed world systems where there are strong political pressures to contain price rises even though new investment needs to be incentivized.

Possibly the greatest vulnerability in an urban context is the supply of transport fuels. The weakness of the low-density settlement is its current complete dependence on oil at a population density below around 30–40 persons/hectare (Levinson and Kumar 1997). There are ways to save oil quickly (IEA, 2005), largely based on increasing vehicle load factors, but there is a mounting recognition of the advantages of diversifying the transport energy vector away from oil.

8.2 Urban public and private sector opportunities and responses

8.2.1 Concepts of sustainable cities and designs

The qualifier 'sustainable' is not exceptional in itself but began to be applied as a term of art in public discourse in the 1980s. The term 'sustainable city' dates from the 1990s. The term 'sustainable development' is usually dated from the World Commission on Environment and Development Report (WCED 1987), but the term appears (undefined) in the Commission's terms of reference relating to earlier usages. The WCED has a chapter on the 'urban challenge' where its concern was principally about issues of the urban poor in rapidly growing large cities of the South. The term sustainable development is now more often used in the narrower context of a need to protect the basic quality of the environment that underpins social and economic capital. For this reason the term 'sustainable cities' is more often associated with civic initiatives in cities of the North, addressing what is perceived as the unsustainable impact of their citizen's lifestyles, especially the generation of large volumes of waste and GHG emissions. It is largely coincident with the earlier idea of an 'ecocity'. Ecocities essentially try to contain their 'ecological footprint' (Andersson 2006; Jabareen 2006; Kenworthy 2006; Pickett et al. 2008). This special focus means that some projects are not necessarily more broadly sustainable, especially as economic entities.

Attempts to achieve an optimal 'sustainable urban system' in new settlements invariably involves some form of spatial organization. This may be provided by a city authority, but it could equally be the covenants imposed by a land developer. The intention is to gain from bringing the various strands of urban activity together into a more integrated whole. For example, reducing urban traffic noise through less need to travel by automobile and the use of quiet road surfaces or electric vehicles enables citizens to keep windows open in summer. This provides the opportunity to avoid mechanical ventilation by recovering the opportunity for effective natural ventilation. 'Sustainable' urban configurations are often expressed in terms of optimal residential densities linked to low-profile effective transport networks. This kind of fairly coarse metric is often employed in zoning regulations. The optimal configuration then seeks to avoid a very high density with highly congested services and very low density automobile-dependent networks. This configuration is expected to induce a stronger sense of community by providing some local retail and commercial space with local interaction, itself reducing the need for automobile travel.

Energy implications of 'sustainable cities' arise naturally from their move away from automobile dependency. But overlaid is often an attempt to exploit the area of the city as a source of renewable energy. At the current state of technology, most 'zero-carbon' developments are essentially *net* zero carbon, and rely on the existence of a market for surplus renewable electricity at some time of the day (and year) and the ability to import electricity at others. The installed PV capacity for the proposed new Masdar city in Abu Dhabi is nearly 200MW within its city boundary of 6 km^2. Wind is less likely to be exploited in a sustainable city, except as an aid to natural ventilation. While wind generators sometimes appear in 'iconic' building structures, drag resistance of the urban surface makes them a less compelling investment than wind generators in more exposed localities managed by the power grid.

Waste is a particular issue for all cities. In eco-city designs such as the Dongtan feasibility study, local waste (in Dongtan's case local agricultural waste) is deployed as a low-grade energy source. But above all the eco-city seeks to reduce the overall level of waste production. Cities of the ancient world produced little non-biodegradable waste, in contrast to the large volume of solid waste that large urban settlements currently need to dispose of. 'Sustainable city' discourses thus focused on reducing waste sent to landfill by providing recycling and incineration facilities. In ironic contrast, the large cities of the developing world already have an informal economy in the periurban areas that picks clean the waste of the formal urban centers. Sustainable cities frequently try to capture waste heat from electricity generation by locating combined heat and power (CHP) sets within the city or encouraging micro-generation. For large urban areas this strategy can only maintain urban air quality if other measures designed to reduce air pollution are successful – which emphasizes the importance of treating issues and systems holistically (Box 8.1).

The 'sustainable city' as conceived at the beginning of the twenty-first century is not without its critics. The sustainability discourse is often so concerned with environmental factors that the local robustness of social and economic capital is unwisely taken for granted. However, examples that have been partially implemented show reductions in final energy consumption against normal benchmarks of 10–15 percent without any substantial changes in lifestyle norms. While sustainable urban form can often require significant capital 'up front' and is thus a serious obstacle under capital constraints, a possibly more fundamental issue that needs to be addressed is *institutional*. Is it possible that stakeholders within an urban context who are used to working independently can find an institutional structure where they can work together in an integrated manner? This may be easily done at the master planning stage, at which broad-brush issues are under the control of a single land-use planner, but to maintain the integrity of the master plan over the years can be a challenge. If the problem can be solved, it would be a major disruptive social technology to conventional urban planning and development and for which proto-type design software is already in existence (Keirstead et al. 2009).

Urban planning measures have the potential to be very powerful methods of integrating urban services in a way that minimizes urban energy use and other ecological impacts. Nonetheless, many of the recent design exercises in 'sustainable cities' are for premium urban centers. The technology may be transferable with further development, but the systems are presently too expensive to treat them as realistic new paradigms for the urban built environment in the immediate decades to come. Possibly the major experience gained is the importance of using a holistic approach to design of the urban environment and that is an approach that can be applied immediately.

Box 8.1 Zero-carbon cities

Planners are exploring new urban paradigms in which urban energy use is not entailed by urban form, but in part defines the urban master plan. Dongtan near Shanghai and Masdar in the Arab Emirates are two recent high-profile design studies that exemplify this.

Both Masdar and Dongtan represent a new development strategy whereby the creation of the city forms its own economic rationale. This reflects the delocalization of some classes of economic activity and hence the ability to bring these together in a desirable location. This is the theory behind the creation of Dubai as a global financial sector from scratch. Less ambitiously, Dongtan was to be a service center outside of Shanghai. Masdar was to be a knowledge center that specialized in renewable energy technologies. The master plans of both complexes are polycentric so that 'quarters' or districts are often defined by a locally dominant economic or social function. Whether the loss of monocentricity is significant is hard to tell. The cellular nature of both development plans enables stable modular growth in uncertain economic times. Both designs have a substantial external energy supply from local renewable resources and so 'to first order' are formally 'zero carbon'. The designs themselves sought to reduce energy demand substantially without affecting service delivery.

Dongtan was designed as a 'zero-carbon' development. The initial settlement study was for 80,000 inhabitants, but with expansion possibilities to over half a million. Its external supply of energy is regional biomass waste from rice production, along with solar power panels. In delivered energy terms, it is an all-electric city. Electric vehicles provide all motorized transport within the city. These can either be conventional battery-powered or fuel-cell powered vehicles that use hydrogen, but the short range is less important in Dongtan because the city is spatially organized to reduce distances for essential travel. Conventional transportation to the outside world has to be parked at the city boundary. The master design features not only low-energy concepts, but also self-sufficiency in water and waste recycling. A particular feature was the exposure of system synergies. For example, by switching to quiet clean electric vehicles, it became possible to revert to natural ventilation and day lighting for housing and offices. The estimated 'eco-footprint' for Dongtan was 1.9 hectares/capita, only slightly above the WWF target (Cherry 2007).

Masdar faces a notably different climate to Dongtan, dominated by cooling load. This demand is reduced by recourse to traditional Arabic architectural approaches that protect building facades and access routes from the sun, with narrow access spaces that still provide daylight penetration into occupied spaces. Like Dongtan, Masdar is a zero-carbon urban development. The scheme's energy supply is provided by a large solar-power energy park. Indeed, Masdar's economic rationale is as an international center in advanced renewable energy technology. Its first districts are already established and it hosts the International Renewable Energy Agency supported by over 130 nations. Industrial zoning places production facilities at the periphery of the neighborhood complex, where more conventional freight transport has access. Novel electric personal transport aims to shuttle residents between centers. As with the Dongtan design, there are no private vehicles within the city. The planned scope of the city is around 6 km^2.

Both Dongtan and Masdar were designed as showcases of a new vision of future master planning focused on low environmental impact, especially of energy. The focus on zero carbon and sustainability reflect the kind of private investment that the developments were intended to attract. The common theme is the value of integrated design in reducing the overall impact of urban processes. It is yet to be demonstrated whether less well-endowed and less well-organized new urban settlements can achieve similar impressive potentials. Given Masdar's initial investments of well above US$20 billion, the sheer magnitude of the investments need for housing some three billion additional urban dwellers in radical new zero-carbon city designs is staggering: well above US$1,000 trillion, or some twenty years of current world GDP (Kluy 2010)!

Whereas the Masdar project has completed at least its first phase, the Dongtan design has been set aside for the moment, although a number of new, perhaps less ambitious, settlements are under construction around Shanghai (Larson 2009). In many ways the Dongtan design exercise has served as a valuable learning experience in 'holistic' master planning and is widely influential. As important as both projects are as showcases and experiments in new thinking and planning, the actual urban development reality is rather one of incremental, continuous change within an existing urban fabric that needs to incorporate the lessons learned from bold 'greenfield' eco-city design experiences.

8.3 Overview of main policy instruments of relevance for urban energy systems

8.3.1 The policy players

Policy instruments that apply to urban settlements are generally exercised down through several layers of government, often as many as five or six. In theory, this plurality is to ensure that administrative powers have an appropriate geographic reach for the issue at hand, and to ensure that destructive competitiveness between settlements does not undermine the quality of specific policy interventions (e.g., Baumol and Oates 1975). Parallel arguments are often deployed to explain the distribution of tax raising, tax collection, and spending powers at different levels of government. The system can be extremely effective, but it can be prone to problems of coordination and regional politics. Unfortunately the almost universal tendency by the highest tiers of government is to decentralize responsibility without fully decentralizing resources. Urban administrations are more often the delivery agent rather than the tax raiser and this frequently leads to accusations by lower tiers of government of 'underfunding' by higher levels. From time to time many of the cities, even in the richest nations, operate close to bankruptcy, which can severely limit their ability to obtain capital for urban energy projects. Generalizations are of course always dangerous and it must be recognized that urban settlement patterns of governance are, in part, an accretion of local and regional history.

The degree to which 'public policy' is a meaningful term in an urban energy context varies greatly from the highly organized urban societies such as Singapore to the current anarchy of Mogadishu. The powers important to energy consumption, such as regulation of construction standards for stationary infrastructure, are to be found at all levels of government. Control of transport provision also occurs at all levels, although land planning, infrastructure standards, and transport can be dealt with by distinct silos within several layers of public administration. Also, landowners, both private and public, often hold important powers through ownership rights that help define the urban form and its physical emergent properties. Houston is possibly the extreme with no state zoning, but relies largely on the operation of private land covenants. Other relevant powers can reside with the public or private bodies that provide the utility services.

Energy consumption is currently an 'emergent' property of a complex urban system. Current governance structures were not designed specifically to manage energy outcomes. In any long-term perspective that embraces a very uncertain future in which energy and related environmental issues become important, it seems likely that greater clarity and effectiveness of governance-relevant structures is highly desirable. The first signs of this need being met are appearing beyond the experiments of the eco-cities. Increasingly, major cities in the world have developed 'energy plans' or 'energy strategies' (e.g. Mayor of London 2004).

8.3.2 The policy instruments

Policy instruments are conveniently arranged in a hierarchy of leverage effects, with land-use planning conceptually at the base and management of infrastructure use at the apex. Sometimes the whole hierarchy is delivered at one stroke, as in a major rapid expansion of an economic zone, and energy optimization or 'zero carbon' has featured in a number of recent expansions (e.g., Masdar (Biello 2008), Dongtan (Normile 2008), and Incheon (Kim and Gallent 1998)).These large enterprises are usually led by a development corporation or similar entity with powers to integrate the various tiers of provision and exceptional access to capital. While impressive in concept, to devise a master plan durable against changes in external circumstances over decades is no mean feat (as the failure to realize the Dongtan project illustrates).

Normally, land-use planning is less ambitious and sets aggregate parameters for zones, such as permitted functions, density, or maximum building height, that guide rather than direct public and private investment. This is the pragmatic solution for the incremental redevelopment of an area. This means that upgrading and refurbishment frequently takes place in patchworks such as '@22' in Barcelona or 'Thames Gateway' in London, usually within the framework of a local or regional government 'master plan'. This approach is potentially very powerful for realizing some of the advantages of economies of scale in low-energy or low-carbon technologies, but there is as yet relatively little experience as to how to exercise it effectively. There are some impressive examples in European towns, such as Malmo and Freiburg, and US towns such as Portland or Davis. These examples influenced recent thinking in urban design, but are apparently not quite impressive enough to induce on their own widespread replication at current (low) energy or 'carbon' prices.

In many jurisdictions the planning authority has the power to apply conditions to new developments, such as mandatory connection to a district-heating scheme. These conditions can be overridden by other conflicting energy-policy objectives. For example, in Europe a new housing development might be required to install a renewable energy source, but energy competition policy would prohibit the imposition of an additional requirement for its exclusive use by housing tenants! Planning control can be effective in eliminating the most excessive energy consumption while giving some certainty to capital investment. However, because so much of the final energy consumption is delivered by instruments at lower levels in the hierarchy, planning control can seldom deliver very low-energy solutions on its own. Where a low-energy solution collides with other environmental requirements, such as appearance or noise, it can militate against it.

The next level 'down' from planning control conventionally contains instruments that relate to the economic framework of the urban settlement. Since the 1980s the trend has been for local services to be provided by the private sector within a regulatory framework. Energy prices are usually regulated (or subsidized) at the national level, and not always in a manner conducive to good outcomes. Removing energy subsidies is theoretically a low-hanging fruit. In practice, the objections of those who benefit from them make such action politically contentious. However, lower tiers of government still have a number of economic instruments at their disposal, especially for transport. Thus, parking charges and more ambitious road-user charging can be used to favor or subsidize mass-transit systems. As these instruments are often redistributive, even when effective, their application can require considerable political skill. The London congestion charge is a case in point (Taylor 2004). More broadly, the price of land and stationary infrastructure is an important factor in all that goes on in urban settlements. Property taxes and taxes on sales can incentivize upgrading of the energy efficiency of the building stock. Where land ownership is heterogeneous, local government can facilitate the roll-out of refurbishment programs, often working with large property developers. Programs of this kind can make up a substantial part of the work of 'energy-service companies' or ESCOMs (Dayton et al. 1998). The lowest tier of policy instruments relates to individual components and their direct use. Construction standards when coupled with standards for energy-control provision have a long history of application. The diffusion of new technologies is *inter alia* influenced by social networks (Fisk 2008) and local networks, of which local government is a part that can accelerate take-up.

The provision of a mass-transport system is an important option when travel densities are sufficiently high to provide savings over private transport. However, success depends, in part, on delivering a service that is universal rather than just for the most disadvantaged. That, in turn, implies a system with sufficient coordination to ensure feasible journey times for a wide range of journeys.

Finally, urban administrations have an important role to play in administration during energy-security events (IEA 2005).

Part III

URBAN POLICY OPPORTUNITIES AND RESPONSES

9

Drivers of urban energy use and main policy leverages

Xuemei Bai, Shobhakar Dhakal, Julia Steinberger, and **Helga Weisz**

9.1 Introduction

This section synthesizes existing knowledge of the main drivers of urban energy use and related policy considerations. Traditionally, comparisons and analyses of energy use and the drivers of differences are carried out at the national level. In comparison, research on the factors that determine urban energy use is still in its early stages, severely hampered by the limited availability of comparable city-level data.

Keeping the above caveats in mind, the factors that determine urban energy use can be classified into a few major groups: *natural environment* (geographic location, climate, and resource endowments), *socioeconomic characteristics* of a city (household characteristics, economic structure and dynamics, demography), *national/international urban function and integration* (i.e., the specific roles different cities play in the national and global division of labor, from production and a consumption perspectives), *urban energy systems characteristics including governance and access* (i.e., the structure and governance of the urban energy supply system and its characteristics), and last, but certainly not least, *urban form* (including the built urban environment, transportation infrastructure, and density and functional integration or separation of urban activities).

These factors do not work in isolation, but rather are linked and exhibit feedback behavior, which prohibits simple linear relations with aggregated energy use. The interaction between the driving factors may change from city to city – moreover, many of the factors are dynamic and path dependent, i.e., are contingent on historical development. There is, however, one factor that underpins all these determinants in a complex and nondeterministic way: the *history* of a city. The location of a city and the initial layout of its urban form are determined historically: witness the difference between sprawling North American cities that developed in the age of the automobile and older, compact European cities that developed their cores in the Middle Ages. Likewise, the economic activities of a city often stem from historical functions, whether as a major harbor, like Cape Town and Rotterdam, an industrial center, like Beijing now and Manchester historically, or a market and exchange center, like London, New York, and Singapore. These historical legacies may have long-term implications on urban energy use. However, there are also cases in which relatively rapid changes in the historical layout and/or the economic role of a city occur. This can be the result of war, natural disasters, or rapid socioeconomic transitions, such as industrialization or deindustrialization. Examples are Tokyo after World War II, Beijing in the past decade as transformed by China's accelerated transition from an agrarian to an industrial society, or many Eastern European cities after the fall of the Iron Curtain in 1989 and the subsequent economic restructuring from a centrally planned toward a market economy.

9.2 Geography, climate, and resource endowments

Climate is an important factor in determining final energy use, especially for heating and cooling demands. Its influence on energy use can be measured through the metrics of heating and cooling degree days, which, in combination with the thermal quality of buildings and settings for indoor temperature, determine energy use. Urban energy demand is, in principle, not markedly different in its climate dependence than that in nonurban settings or national averages, but it is structured by the influence of other variables, such as urban form (e.g., higher settlement densities lead to smaller per capita residential floor areas), access to specific heating fuels, or income (e.g., more affluent urban households use more air conditioning), that can amplify or dampen the effect of climate variations on urban energy demand.

National studies illustrate the quantitative impact of climate variables on energy demand. For example, Schipper (2004) reports differences in space-heating energy use (measured as useful energy) normalized to heating degree days and square meters living space for seven industrial countries. This analysis reveals substantial ranges from 50 kJ/m²/degree-day for Australia to 250 kJ/m²/degree-day for the United States in the early 1970s, and from 60 (Australia) to 160 kJ/m²/degree-day for Germany in the mid-1990s. Assuming a residential floor space of 100 m², a difference of 1500 heating degree-days, which is characteristic between northern (Denmark) and southern (Greece) Europe, translates into a variation in residential energy demand between 9 and 24 GJ, which is significant compared to a typical European household residential energy use of some 60 GJ, but nonetheless only constitutes between 9 percent and 24 percent of the typical 100 GJ/capita Western European total urban final energy use. Conversely, little is known on the differences in the demand for thermal comfort as reflected in ambient indoor temperatures. A case study carried out for Metro Manila indicated that people in the highest income brackets have much lower indoor room temperature setting preferences, which leads to an increased air-conditioning demand (Sahakian and Steinberger 2010).

The relationship between climate and urban energy use is a two-way street: climate not only influences urban energy demand, but urban areas also influence their local climate through the 'urban heat island' effect (see Chapter 7). This effect can reduce the heat demand during winter, but also enhance the need for cooling in the summer, especially in warm and humid climates. Studies show increases in the summer time cooling load in tropical and midlatitude cities (Dhakal et al, 2003). A series of studies on California show that a 0.5°C increase in temperature causes a 1.5–3 percent increase in peak electricity demand (Akbari et al. 1997).

To a certain extent cities inherit the resource dependencies of their respective countries, which explains, for instance, the continued use of coal in urban areas in countries endowed with large coal resources. The connection to national energy systems and their dependence on the resource base is especially pronounced for power generation, since cities often draw electricity from the national grid. In some cases, urban power plants are designed to use local resources, such as hydropower, geothermal, or wastes, but these potential resources are usually extremely limited in urban areas and provide only a small contribution to the high energy demand associated with high urban population and income densities. On the distribution and end-use side, district heating and cooling infrastructures, which allow reaping of scale and scope economies, cogeneration, and energy-efficient 'cascading' schemes, are specific urban-efficiency assets, but only economically possible when the density of demand is above a threshold that warrants the investment.

9.3 Socioeconomic characteristics

The positive correlation between income and (final) energy use is long established in the traditional energy literature, especially for analyses at the national level. For the household level, correlations between income and energy use have been shown for the Netherlands (Vringer and Blok 1995), India (Pachauri and Spreng 2002), Brazilian cities (Cohen et al. 2005), Denmark (Wier et al. 2001), and Japan (Lenzen et al. 2006), with similar results for GHG emissions in Australia (Dey et al. 2007) and CO_2 emissions in the United States (Weber and Matthews 2008). For Sydney, Lenzen et al. (2004) showed that urban household energy increases with household expenditure, and that most of this increase results from the energy embodied by goods and services, since direct final energy use, in contrast, increases only slowly with expenditure (albeit from high baseline levels).

Based on a production approach, urban per capita energy use is very often lower than nonurban energy use or the national average, particularly for postindustrial, service-sector oriented cities in the OECD countries (see Chapter 5; Brown et al. 2008; Parshall et al. 2010).

Figures 9.1 and 9.2 show the urban income–energy relationship from a production perspective. The GRP/resident is plotted in a cross-sectional analysis against energy use for a sample of Chinese cities (Figure 9.1). Figure 9.2 complements the Chinese cross-sectional analysis by a longitudinal analysis for six megacities. For both cases, income and energy increase together, albeit along distinctly different trajectories, which illustrates *path dependency*. Income is therefore far from the sole determinant of the level of energy use: for instance, Beijing and Shanghai have a higher average energy use than Tokyo, despite a lower per capita income.

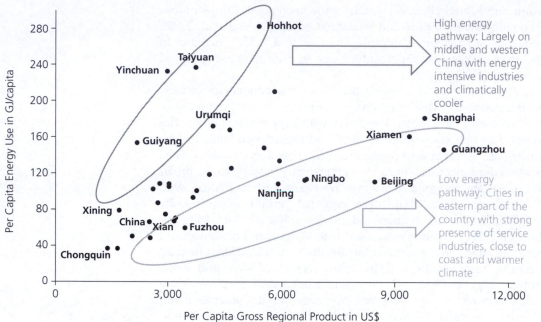

Figure 9.1 Per capita energy use (GJ/capita) versus income for a sample of Chinese cities for 2006, illustrating path dependency

Source: Dhakal (2009) (per capita GRP is expressed in US$2006 calculated using market exchange rates (MER))

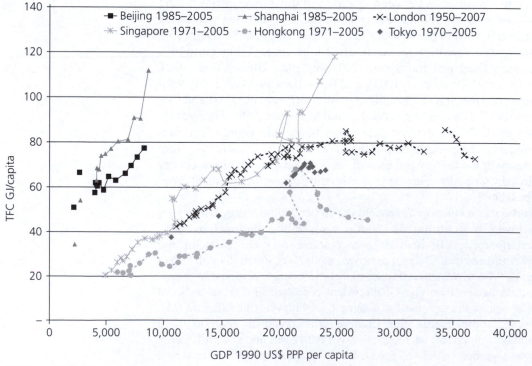

Figure 9.2 Longitudinal trends in final energy (GJ) versus income (at PPP, in Int.$ 1990)[1] per capita for six megacities. Note the path-dependent behavior

Source: Schulz (2010b)

In addition to income, demographic factors play a role in determining urban energy use (Liu et al. 2003; O'Neill et al. 2010). For instance, studies suggest that household size, that is, the number of people living in one household, plays a role in energy use: above two people per household, economies of scale can reduce the energy use per capita. This phenomenon is observed in India (Pachauri 2004), Sydney (Lenzen et al. 2004), the United States (Weber and Matthews 2008), and Denmark and Brazil (Lenzen et al. 2006). In Japan, in contrast, larger household sizes correlate with slightly larger energy use (Lenzen et al. 2006). Urban populations often have significantly smaller household sizes than rural populations because of smaller families and a larger generation gap, as well as smaller dwellings, and so shelter for extended families or many generations under the same roof is less likely.

The evidence for age dependence is mixed. In Sydney, increasing age is correlated with a citizen's higher residential, but lower transportation energy use (Lenzen et al. 2004). Larivière and Lafrance (1999) found a positive correlation between residential electricity use with age for Canadian cities. At this point, not enough is known regarding the influence of age to make any general statement, much less predictions, applicable to cities with very diverse age pyramids that range from young and growing, to old and declining populations.

9.4 Role of the city in the national or global economy

A city's function in regional, national, and international economies has a strong bearing on its energy signature when measured from a production perspective. In the extreme case of Singapore (Box 9.1), a major center for oil-refining and petrochemical production and a major international transport hub, the energy use associated with international trade in oil products, shipping, and air transport (usually subsumed[2] under 'apparent consumption' of the city's primary energy use) is four times larger than the direct primary energy use of Singapore and more than eight times larger than the final energy use of the city.

That urban areas are usually in an intense process of energy exchange (imported and exported) with surrounding markets is again shown dramatically in Figure 9.4 for Tokyo in terms of CO_2 emissions. Emissions attributable to the direct and indirect energy and resource uses of Tokyo ('inflows') are balanced with the final consumption categories of these inputs as well as exports ('outflows'). Embodied energy and emission flows have gained increasing importance, comprising some 80 percent of Tokyo's 'inflows' and still slightly half of its 'outflows,' which illustrates Tokyo's embeddedness in the global economy. The Tokyo example also indicates the importance of energy and emissions embodied in maintaining and expanding the physical infrastructures and capital goods (reported under capital formation in Figure 9.4). Generally, private households only account for a small fraction of total capital formation (dominated by government, industry, and commerce). A consumption-based accounting that only uses

Box 9.1 Singapore: the importance of trade[3]

The case of Singapore illustrates the intricacies of energy (and emissions) accounting in trade-oriented cities that import primary energy, such as crude oil, re-export processed energy (fuels), energy-intensive products (petrochemicals), refuel ships and aircraft (bunker fuels), and import and export numerous other products and services that all 'embody' energy (see Figure 9.3). In terms of energy or CO_2 emission accounting, this extreme example amply illustrates the limitations of applying current inventory methodologies developed for national applications to the extremely open economies of cities. New, internationally agreed accounting standards are needed, as otherwise the risk of either misinforming policy or drawing arbitrary system boundaries is significant. There is a risk of 'defining away' energy use and emissions (e.g., international bunker fuels for aircraft and ships) associated with the inherent functioning of spatially defined entities (cities, city states, even small national open economies) whose interdependencies and energy/emissions integration into the international economy provide for their very *raison d'être* and therefore need to be included in energy and GHG emission inventories.

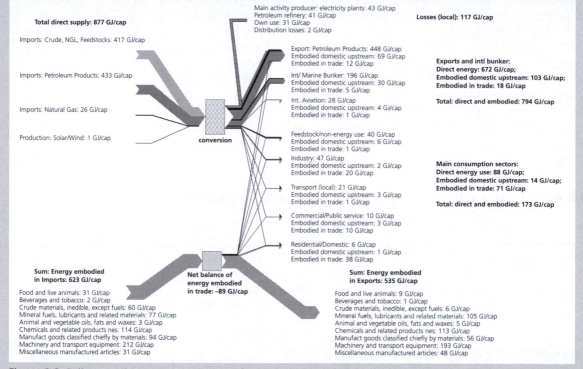

Figure 9.3 Full per capita energy accounting for both direct and embodied energy flows of a large urban trade city, Singapore (in GJ/capita). Domestic direct and embodied energy use is 173 GJ/capita, but is dwarfed by the total energy imports to the city of 1490 GJ/capita. Total energy re-exports (direct and embodied) are 1225 GJ/capita

Source: Schulz (2007, 2010a)

household expenditures (as frequently done) therefore misses these important embodied energy and emission flows.

The thirty-five largest cities in China (China's key industrialization and economic drivers) are responsible for 40 percent of the nation's GDP and contribute overproportionally to national commercial energy use (Dhakal 2009). Cities often specialize in certain types of manufacturing, commercial, or administrative functions. Some urban areas are also large transport hubs, such as London for air transit, or

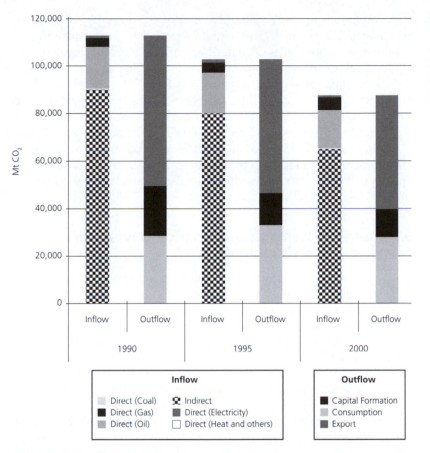

Figure 9.4 CO_2 balance of Tokyo 1990, 1995, and 2000 using I–O analyses, in million tons CO_2
Source: Draft estimates by Shinji Kaneko and Shobhakar Dhakal (2012)

Cape Town and Rotterdam for shipping, that adds significantly to urban energy use, and is too often omitted from urban energy and GHG accounts. For instance, London's twin functions as a major international airport hub and as a global city result in an energy use from air transport that corresponds to one-third of London's total (direct) final energy use (Mayor of London 2004).

A service-based economy can generate the same income with less energy than an economy based on the production of goods, which is one reason city per capita energy use in advanced, service-oriented economies is lower than national averages. This is also why Shanghai and Beijing have higher energy use per capita than Tokyo (see Figure 9.2 above), despite their lower GDP/capita.

If the economic activities located within a city determine its local energy use, its economic transactions with other areas entail energy use in those areas. Any product or service bought or sold entails energy use, and for service-oriented cities it may well be that the energy used, indirectly, through their economic transactions is larger than the energy used locally by their services industry. This phenomenon was shown at the level of urban household expenditures: rich households consume more energy indirectly than they do on housing, utilities, and local transit (Lenzen et al. 2004).

In addition to economic globalization, cultural globalization encourages urban upper and middle classes to adopt consumption patterns from global elites. Globalization-influenced urban development tends to favor private automobile-based individual transport modes and suburban sprawl for those who can afford it. Foreign direct investments (FDIs) and trade agreements affect the location and technology of manufacturing and commercial activities and labor reorganization (Romero Lankao et al. 2005). In China, individual cities compete with each other to attract FDIs and compromise their local environmental conditions and tax policies (Dhakal and Schipper 2005). This type of intra-national competition also occurs in other countries, such as Vietnam and India.

9.5 Energy systems characteristics: governance, access, and cogeneration

The organization of energy markets and their controls at the urban level also influence urban energy use. Alternative organizational forms, such as state or municipal monopolies, cartels, or free-markets, impact access, affordability, and the possibility of implementing energy-saving policies. Localized energy monopolies may work closely with urban governments to further local policies, whereas free-market structures often challenge the enactment of environmental or social policies, such as renewable mandates, or the possibility of performance contracting. New York City requires (because of energy security and reliability concerns) 80 percent of electricity-generating capacity to be located within its territory; this means that the ability to influence the energy system is different to that in other cases. Vienna city owns its respective electricity, gas, and district-heating utility companies, and thus may have greater influence compared to cities with completely privatized and deregulated utilities. In Chinese cities, where energy companies are state-owned enterprises, the city government policies can exert strong influence on the suppliers, albeit less on the energy demand side. Many industrialized cities have put in place City Climate Actions Plans, which are expected to reduce or dampen energy use or promote shifts to renewables in the coming decades, but their success will depend on the links between city government and local energy providers. In many cities across the world, the local government is hardly able to influence the energy-supply side (because of jurisdictional and capacity limits), but may be in a position to address demand-side energy issues.

In developing countries, urban populations generally have higher levels of access to commercial energy forms than rural populations. This affects the efficiency and the intensity of the environmental impacts of energy use (Pachauri 2004; Pachauri and Jiang 2008): rural populations consume (often self-collected) fuels such as fuel wood, biomass, and coal; urban populations consume commercial and cleaner energy forms: electricity, oil, and gas. Owing to the low level of efficiency of biomass use, the quantity of primary energy use per capita may be similar in urban and rural settings (Pachauri and Jiang 2008), but the different fuel structure in urban, higher income settings provides for much higher levels of energy service provision. In this sense, urban populations benefit from the high efficiency of energy-service delivery of modern fuels and distribution systems, such as electricity, gas, or bottled LPG. Access to commercial energy is much less an issue in industrialized or industrializing countries, which already have electrification levels at 100 percent (IEA 2002) and where gas-distribution networks connect a majority of urban households.

Many European countries also have a long tradition of urban district heating (and more recently of district cooling) networks that either use district heating plants or CHP energy systems. CHPs, in particular, offer potential energy-efficiency gains as waste heat from electricity generation can be used for low- and medium-temperature heat demands in urban areas, with steam-driven chillers that also provide cooling energy. Traditionally, such centralized systems are capital intensive and only economic in higher density urban settings that provide for sufficiently high demand loads to warrant the investments. The recent advance in more decentralized energy solutions, including microgrids, allows such systems to be extended to lower density urban settings. Typically, cities with significant energy cogeneration have primary energy needs that can be 10–20 percent lower compared to systems in which all heat demands are provided by separate, individual boilers or furnaces.

A key issue for the improved efficiency of urban energy systems is therefore an optimal matching between the various energy-demand categories and forms to energy-conversion processes and flows, usually achieved by exergy analysis (see Box 9.2).

Box 9.2 Urban exergy analysis: efficiency – how far to go?[4]

An analysis of the efficiency of urban energy systems is far from a trivial task, but it is fundamental to identify options and priorities for improved efficiency in energy use. With respect to the system boundaries of the analysis, should the analysis extend to final energy (the usual level of market transactions in the energy field), to the level of useful energy, or to energy services? Should only simply energy outputs–inputs relationships be considered in defining efficiency (referred to as first law analysis in the literature, after the first law of thermodynamics) or the analysis be extended to consider quality differences in energy forms (which energy form is most adequate for delivering a particular task) and efficiency, not in absolute terms (as in first law analysis), but in relation to what thermodynamically represents an upper bound of energy conversion efficiency (as no conversion process that operates under real-world conditions can achieve 100 percent efficiency)? The latter concept is referred to in the literature as second law (after the second law of thermodynamics), or *exergy* analysis (e.g., Rosen 1992).

The literature (e.g., Nakicenovic et al. 1990; Gilli et al. 1995) identifies the value of both types of analyses (first and second law analysis), but also concludes that second law analysis enables us to extend the system boundaries to include also energy *service efficiency* (which cannot be captured in first law analysis as it lacks a common energy denominator) and important quality characteristics of different energy forms and their adequacy to deliver a particular energy service. Therefore, an illustration of the value of exergy analysis to assess the efficiency of urban energy systems is provided here using the example of Vienna, which is compared to a few fast-track European urban-exergy analyses obtained from various research groups.

The energy system of the city of Vienna is characterized by a number of unique features. First is that the city generates much of its electricity needs within the city itself allowing the use of resulting waste heat through a district-heating network (recently also extended to a district-cooling network). As a result, the corresponding first law efficiencies of Vienna's energy system are very high: 85 percent of secondary energy is delivered as final energy and about 50 percent can be used as useful energy to provide the energy service needs of the city (see Figure 9.5). The impact of cogeneration on the city's energy needs is also noticeable: without cogeneration Vienna's secondary energy use would be some 13 percent higher. The high first law efficiencies suggest limited improvement potentials. However, this is not the case as revealed by a second law analysis of Vienna's energy system, which shows the efficiency between secondary and useful exergy is only some 17 percent. This suggests significant improvement potentials, for example via heat-cascading schemes that better match the exergetic quality of energy carriers with the required temperature regime of energy end-uses.

This assessment obtained the results of similar exergy analyses for Geneva, Switzerland (Giradin and Favrat 2010), the Swedish city of Malmo, and London (Fisk 2010). The results of the comparison in terms of the efficiency of useful exergy to that of secondary and primary exergy are summarized in Table 9.1.

Table 9.1 Comparison of the efficiency of useful exergy to that of secondary and primary exergy

	Useful exergy as % of:	
	secondary	primary
Geneva (CH)	23.2	15.5
Vienna (A)	17.2	
Malmö (S)	21.2	12.7
London (UK)	11.3	6.2
trad. Mexican village	5.7	

The results confirm earlier conclusions that, thermodynamically, urban energy systems could, in theory, be improved vastly, perhaps by as much as a factor of 20 (a similar order of magnitude as suggested by Nakicenovic et al. (1990) for OECD countries), and thus leave ample opportunity to realize feasible measures under real-world conditions and constraints that might deliver an improvement by at least a factor of 2. That modern urban energy systems are – despite their comparatively low exergy efficiencies – vastly more efficient (by a factor of 2–4) than traditional rural energy systems is also shown in the Masera and Dutt (1991) analysis of a traditional Mexican village with 2,400 inhabitants using mostly preindustrial energy forms and conversion technologies (draft animals and fuelwood) for the provision of their energy services yielding an exergetic efficiency of only some 6 percent, compared to 11–23 percent for modern urban energy systems and uses.

Figure 9.5 Energy and exergy flows in the City of Vienna in 2007 between secondary and useful energy/exergy. (See color plate 11)
Sources: Energie Wien (2009) (approximate) exergy efficiencies based on Gilli et al. (1996)

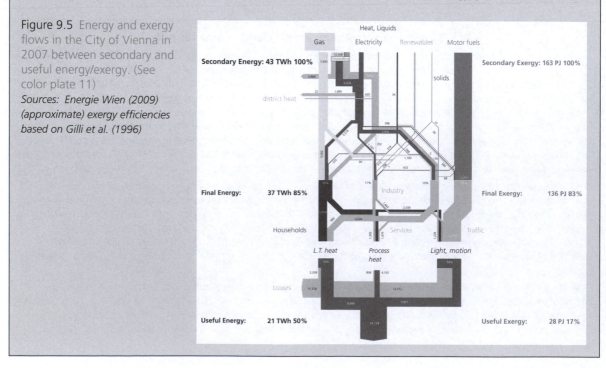

9.6 The urban form: the built urban environment and its functions[5]

9.6.1 The built environment

The built urban environment comprises the totality of the urban building stock: residential, commercial, administrative, and industrial buildings, their thermal quality and spatial distribution. It also includes built urban infrastructures for transport, energy, water, and sewage. This environment is one of the key components for understanding the special characteristics of urban energy use as compared to rural, economy-wide, or global patterns. The unique concentration and overall scale of the built urban environment allow both economies of scale and economies of scope to occur, and thus provide options for energy-efficiency gains.

9.6.1.1 Building design

The design and thermal integrity (e.g., insulation levels) of buildings are essential for the amount of energy intensity (energy/m²) needed for heating and cooling. Reducing the energy associated with heating has been a strong focus in northern European countries, but mid-latitude countries have to attempt a design a balance between heating and cooling energy demands. In many cases, newer buildings have better thermal standards, but in some cases they are poorly adapted to their climate (e.g., European- and US-style villas and apartment buildings in tropical climates, which do not have adequate shade and ventilation). Old buildings may suffer from lack of renovations, or renovations that do not apply the best possible standards. The influence of building technology on the energy used for space heating is huge: a Passivhaus standard requires that energy use for space heating be no more than 15 kWh/m² floor area per year; for low-energy houses the corresponding number is around 50 kWh/m²/yr, whereas poor thermal insulation may cause energy use for space heating of 200–400 kWh/m²/yr in mid-European latitudes.[6]

The energy involved in the maintenance and replacement of components over a building's life should also be taken into account in assessing the energy performance of a building. For a 50+ year lifetime of office buildings, the embodied energy in construction materials plus the energy needed for decommissioning is estimated to range from 2.5 to 5 years of the building's lifetime operational energy use (Cole and Kernan 1996; Scheuer et al. 2003; Treberspurg 2005), with a typical value of embodied energy being between 5 percent and 10 percent of direct, operational energy needs of buildings. Including single and multifamily houses somewhat expands this range. The detailed literature review of Sartori and Hestnes (2007) reports a range from 4 to 15 percent embodied energy in total lifecycle energy use of buildings. Only in extremely low energy-use buildings (e.g., Passivhaus-standard or even below), with their extremely low operational energy use, does embodied energy play a somewhat greater role, reaching between 25 percent (Sartori and Hestnes 2007) and 29 percent (Treberspurg 2005): typical values of 20–30 kWh/m² building floor area/year of embodied energy compare to 50–60 kWh/m² building floor area/year for operational energy (heating plus electricity).

9.6.1.2 Type of buildings and uses

Next to the energy characteristics of an individual building, also the mix of building types and their density are important determinants of urban energy use.

The specificities of the urban built environment are usually a large existing stock, which requires renovation and maintenance, and new buildings in growing cities. The improvement in building stock to lower heating and cooling demands is counterbalanced by the increase in surfaces necessary to house new populations in growing cities, along with the demand of inhabitants for larger and larger apartments – even as the average household size decreases. Residential floor space per capita is known to be strongly correlated with income (e.g. Schipper 2004; Hu et al. 2010). National averages in industrial countries range from 30 m²/person in Japan to 50 in Canada, 55 in Norway, and 80 in the United States (Schipper 2004; US DOE 2005). Typically, urban residential floor space per capita is lower than the national averages (to a degree counterbalanced by smaller household size), particularly for high-density cities with their corresponding high land and dwelling prices, but comprehensive statistics are lacking. For urban China, Hu et al. (2010) estimate 5 m²/person in 1990 and approximately 25 m²/person in 2007.

Newton et al. (2000) evaluated and modeled the energy performance of two 'typical' dwelling types – detached houses and apartments – across a range of climatic zones in Australia. Two main conclusions were drawn: (1) annual heating and cooling energy and embodied energy per unit area were similar for apartments and detached houses; (2) per person, however, the lifecycle energy use of apartments was significantly less (10–30 percent) than that of detached houses in all circumstances, because the area occupied per person was much less. Norman et al. (2006) used a lifecycle analysis approach to assess residential energy use and GHG emissions, contrasting 'typical' inner-urban, high-density and outer-urban, low-density residential developments in Toronto. They found that that the energy embodied in the buildings themselves was 1.5 times higher in low-density areas than that in high-density areas on a per capita basis, but was 1.25 times higher in high-density areas than that in low-density areas on a per unit living area basis. Salat and Morterol (2006) compared eighteenth-century, nineteenth-century, and modernist urban areas in Paris, assessing five factors in relation to CO_2 emissions for heating: (1) the efficiency of urban form in relation to compactness; (2) a building's envelope performance; (3) heating equipment type, age, and efficiency; (4) inhabitant behavior; and (5) type of energy used. Salat and Morterol (2006) asserted that an efficiency improvement factor of up to 20 could be achieved from the worst-performing to the best-performing urban morphology by taking these five factors into account. Salat and Guesne (2008) investigated a greater range of morphologies in Paris and found that when considering heating energy, the less dense the area, the greater the energy required for heating (see also Ratti et al. 2005).

9.6.2 Urban form and functions

Urbanization patterns affect the extent and location of urban activities and impact the accompanying choice of infrastructures. Newton et al. (2000) summarized key alternative urban forms or 'archetypal urban geometries,' namely the dispersed city, the compact city, the edge city, the corridor city, and the fringe city. The merits of dispersed and compact cities ('suburban spread' versus 'urban densification') have been debated since the 19th century and a strong divide exists between the 'decentrist' (the dispersed city model) and 'centrist' (the compact city model) advocates (Brehny 1986).

Nonetheless, one the most important characteristic of cities is density. Overall, a certain density threshold is the most important necessary (although not sufficient) condition to allow efficient and economically viable public transit (see Chapter 10). In addition, in a dense environment distribution networks are shorter, infrastructure is more compact, and district-heating and -cooling systems become feasible. Unconventional energy sources, such as sewage and waste heat, are also more accessible. High density may thus help curb urban energy use (Rickaby 1991; Banister 1992; Ewing and Cervero 2001; Holden and Norland 2005).

Most importantly, a compact city brings the location of urban activities closer. In the context of transportation, from cross-city comparisons it is well established that higher urban densities are associated with less automobile dependency and thus less transport energy demand per capita (Newman and Kenworthy 1989; Kenworthy and Newman 1990; Newman and Kenworthy 1991; Brown et al. 2008; Kennedy et al. 2009). Intra-city studies for Sydney (Lenzen et al. 2004), Toronto (VandeWeghe and Kennedy 2007), and New Jersey (Andrews 2008) also show that denser neighborhoods have lower per capita transportation energy needs. As a result, in many low-density cities, per capita energy use has grown at approximately the same rate as that in sub-urban areas (sprawl) (Baynes and Bai 2009).

In many of the less-compact cities, transportation[7] by automobile is the biggest contributor to energy use (Newman and Kenworthy 1999). The data suggest that cities with a density of 30–40 people/ha or greater developed a less automobile-based urban transport system with typical density thresholds for viable public transport systems given as above 50–100 people/ha (see Chapter 10). On average, residents who live at a distance of 15 km from an urban center use more than twice the transport energy compared to residents living 5 km from the center (Stead and Williams 2000). Nijkamp and Rienstra (1996) note that the private automobile has brought low-density living within the reach of large groups of upper and lower middle-class families. Moreover, correlations between automobile ownership and income suggest that more affluent automobile owners have a higher propensity to travel longer distances by energy-consumptive modes (Banister et al. 1997).

Diversity of function may also play a role in managing urban transport demand (Cervero and Kockelman 1997). When strict zoning is enforced so that residential areas are separated from commercial, education, services, and work areas, private transportation is maximized. Mixed land uses and concepts of self-containment are important in reducing energy use in transport. Nevertheless, local jobs and local facilities must be suitable for local residents, otherwise long-distance, energy-intensive movements will continue (Banister et al. 1997). This coordination of land-use and transportation policies is termed transit-oriented development. The idea of location efficiency emphasizes the accessibility of opportunities, rather than how mobile one must be to find them (Doi et al. 2008); this is a central concept in recent approaches to transit-oriented development and other forms of sustainable urban development.

Also, urban density is an indicator of *potential* energy savings, especially in transportation. If infrastructure is inadequate to support the volume of traffic flow, the resulting congestion can lead to higher energy use, even in high-density, built-up areas. For energy efficiency potentials of urban densities to be realized, a chain of interdependent, appropriate infrastructure, technical, and consumption decisions must be made. The correct level of public transit infrastructure requires large up-front investment and maintenance, from light rail to subways, trams, or dedicated bus routes. Adopting public transit also requires appropriate consumer behavior. In many North American cities, public transit is associated with lower economic status, and thus avoided by most people who can afford to drive, which reinforces the initial perception. A contrary example is Tokyo where the per capita energy use is smaller than in many East Asian megacities; one of the key reasons for this is the efficient rail-based public transport in Tokyo (Dhakal 2004, 2009).

Another important energy implication of the urban form is the choice of urban energy-supply systems. District-heating and -cooling infrastructures, which allow large economies of scale and efficiency gains through cogeneration, are only possible when the density of demand is high enough to warrant the capital-intensive investment, unless such systems are mandated (and costs added to land prices). Compact urban form may also play a role in the energy used for buildings. Apartment buildings generate economies of scale compared to single-family homes, but apartment buildings may compromise decentralized low-energy design practices, such as natural lighting, ventilation, and decentralized use of PVs. Another important influence of density is at the personal consumption level. Apartment size per person tends to decrease with population density (with Hong Kong and Manhattan representing extreme examples). Effectively, the high competition for central urban space creates rents that contract floor space. However, in cities without sufficient low-rental housing, even the smallest apartments can be out of reach for the poorer populations, who are forced to live in distant suburbs with poor transit connections. In many cities, suburbanization is also caused by industrial relocation from urban cores and the unplanned settlement of migrants and urban poor in the urban periphery.

More compact cities, however, may require special management to avoid the ill-effects of congestion and higher concentrations of local pollution (e.g., see Jenks et al. 1996). Urban heat island effects, for instance, may be exacerbated in dense urban cores. There may be a trade-off between the transport energy savings achieved with higher urban density versus the higher energy use of high-rise buildings. There are also trade-offs between urban density, dwelling type, block size, and the ecosystem services provided by vegetation. Both theoretically and empirically, it is by no means clear that there is an ideal urban form and morphology that can maximize energy performance and satisfy all other sustainability criteria.

9.7 Relative importance of the drivers of urban energy

No study so far has investigated the relative importance of all the factors known to influence urban energy use as described above. Existent approaches rather contrast energy and/or CO_2 emissions with such macro-drivers as population, income, and technology, and thus follow the classic IPAT decomposition approach.[8] Such decomposition analysis has, for example, been carried out for several Chinese cities (Dhakal 2009), where the relative changes in urban CO_2 emissions are decomposed into the factors population change, income change (measured as GDP/capita), and two technology factors: the carbon intensity of the energy system (measured as CO_2 emissions per unit of primary energy demand) and energy intensity (measured as primary energy demand per unit of GDP) for several periods of time. Although the relative contribution of these factors varies across cities and time periods, overall income is shown to be the most important driving factor for increases in carbon emissions (by far outpacing population growth), and improvements in energy efficiency to be the most important counterbalancing factor. The net result is, in all cases, an increase in carbon emissions, which indicates that economic growth has, to date, outpaced technology and efficiency gains (Figure 9.6).

Earlier work by Dhakal and Hanaki (2002) and Dhakal et al. (2003) for Tokyo using 1970–98 data and for Seoul using 1990–7 data also shows that the income effect was primarily responsible for the majority of energy-related CO_2 emissions growth in Tokyo and Seoul in their respective high growth periods, that is, 1970–90 for Tokyo and 1990–7 for Seoul. The analysis also showed that, despite an economic recession, energy-related CO_2 emissions continued to grow in Tokyo in 1990–8, largely because of a drastic decline in the energy-intensity improvement rate (often observed in periods of economic growth stagnation or

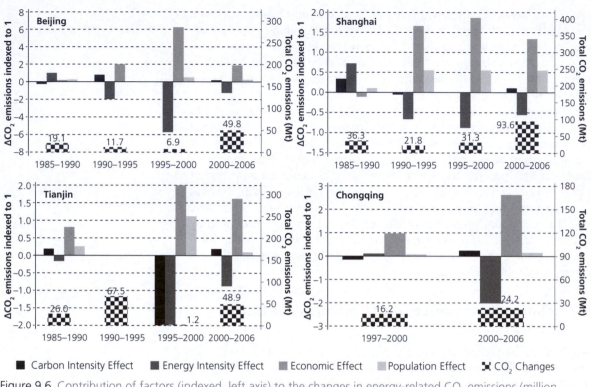

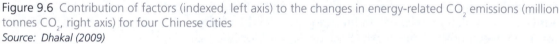

Figure 9.6 Contribution of factors (indexed, left axis) to the changes in energy-related CO_2 emissions (million tonnes CO_2, right axis) for four Chinese cities
Source: Dhakal (2009)

recession caused by the slower rate of capital turnover and hence the slowing introduction rate of more energy-efficient technologies and practices).

Notes

1 For comparison: per capita GDP (in PPP terms) in 2005 and in Int.$2005 are: Beijing: 9,238, Hong Kong: 3,4574, London: 5,3145, Shanghai: 9,584, Singapore: 2,9810, and Tokyo: 3,3714. (Note that a change in base year for the PPP metric changes the relative position of urban incomes in a non-proportional way.)

2 International bunker fuels are an important exception that, by simple definition, are excluded in national energy-use balances and the resulting emission inventories.

3 Author: Niels B. Schulz.

4 Authors: Arnulf Grubler and David Fisk.

5 A working paper on urban form and morphology contains a more extended discussion and is available online at: www.globalenergyassessment.org.

6 See http://energieberatung.ibs-hlk.de/ebenev_begr.htm.

7 For a more in-depth discussion of transport energy use and its drivers, see also Chapter 9 of GEA.

8 IPAT: Impacts = Population × Affluence × Technology. For a history and discussion of the concept, see Chertow (2000).

10

Transport systems

Gerd Sammer

10.1 Urban travel behaviour

To a large extent, travelling and mobility are means to an end; they are necessary to enable people to fulfil essential functions, i.e. living, working, gaining education, acquiring necessary supplies, and relaxing, in the most suitable places. The situation is similar for the production of goods and services: division of labour means that production and manufacturing can take place at different sites, which improves the efficiency of production. This causes travel demand, which is measured in different ways depending on the reasons for travel:

- Frequency of trip making. This metric is a basic indicator of the degree of mobility and concurrent degrees of economic development and lifestyles. In urban settings this figure ranges from 2.2 trips per day and person in developing countries (Padam and Singh 2001) to up to 4.0 trips in industrialized countries (Hu and Reuscher 2004; Sammer 2008). It is expected that in future the frequency of trip making will rise slightly, particularly in developing countries. Generally the frequency of trips is higher in cities than in rural areas because of higher degrees of specialization and division of labour and the higher number of attractive destinations with good accessibility, which are characteristic for urban areas.
- Distance travelled per day. On the one hand this metric reflects the modes of transport used and on the other hand the spatial structure and settlement density. The lower the settlement density and the more car-oriented an area, the longer the distances travelled per day. Globally these distances vary considerably, from 10 to 60 km per person per day in developing and industrialized countries. In rural areas of developed countries the average distance travelled per day is higher than in cities. In developing countries the opposite is true: in rural areas distances travelled are generally very low (significantly below 10 km per person) and are significantly higher in urban areas due to both higher urban incomes as well as better availability of urban infrastructures and transport options. There is a close positive correlation between the distance travelled per day and transport energy use.

- Travel time budget per day. This metric is an indication of the time spent travelling. For long periods this figure has been comparatively stable, with a slight tendency to increase; it is currently nearly up to 70 minutes per person/day and 90 minutes per mobile person/day in countries with high motorization and car orientation (Hu and Reuscher 2004; Joly 2004). Assuming that in the long run working hours will decrease due to continued productivity gains, one can expect this figure to increase somewhat in future.

Traffic surveys help to determine the travel behaviour which is typical of various sections of the population and for different settings. Caution is however advised when comparing different mobility surveys due to differences in survey coverage and methods (e.g. does the survey include non-motorized modes?) and differences in sampled populations (commuters to work versus total population) and other methodological intricacies. Therefore, there remain serious data gaps for comparable and up-to-date mobility surveys across a wide range of settlements and with comprehensive geographical coverage. The section below summarizes the current state of knowledge with special emphasis on survey comparability rather than recentness of survey date. Despite important data limitations, nonetheless some robust generic patterns can be discerned.

10.1.1 Modal split of cities

The modal split is an expression for the shares of different modes of transport in the overall travel demand – usually measured as shares in total number of trips performed. It is a reasonable proxy for the evaluation of the environmental soundness of an urban transport system and its associated energy use: higher shares of so-called 'ecomobility' transport modes (including non-motorized modes like walking and cycling as well as public transport) and corresponding smaller shares of private motorized transport imply a more environmentally friendly and energy-efficient transport system. The modal split can be conveniently depicted in a triangular diagram (Figures 10.1–10.2) plotting the respective shares of non-motorized modes (walking and cycling), public transport, and private motorized transport (cars, two-wheelers, etc.) respectively. In cities or towns with a modal-split point close to the centre of the triangle the three modes of transport have fairly similar shares. Public transport dominates in the case of modal-split points close to the top of the triangle (e.g. New Delhi in 1994 in Figure 10.1); private motorized transport is dominant in the lower right-hand corner (e.g. Chicago in 2001 in Figure 10.1) and non-motorized transport in the lower left-hand corner of the triangle (e.g. Lucknow in 1964 in Figure 10.1). This representation enables both to display cross-sectional as well as longitudinal observations.

Figure 10.1 provides a comparison of the average modal split in cities and towns on all continents (Kenworthy and Laube 2001; Padam and Singh 2001; Zhou and Sperling 2001). Private motorized transport has a very high share between 88 and 79 per cent in cities in the USA and Australia, where ecomobility has only a very low share of 12 to 21 per cent. This means that in these cities and towns transport systems depend strongly on private cars and thus on fossil fuels. Cities and towns in Western Europe and well-developed Asian economies are roughly in the centre of the diagram. There motorized private transport holds a share between 50 and 36 per cent, i.e. transport systems are less car oriented. Ecomobility holds a remarkably high share of 10 to 50 per cent. In the cities and towns of China private motorized transport has thus far been fairly insignificant with a share of only more than 10 per cent, but is growing rapidly in importance. Figure 10.1 can be interpreted as follows: with growing economic development and increasing incomes, the share of private motorized transport increases while the share of ecomobility decreases. This development is well illustrated by the longitudinal time trends of the modal split of the city of Lucknow in India over the period 1964 to 1998. The USA and Australia seem to have reached saturation in this development; Western Europe and Asia have still some potential to develop further in this direction, unless they take countermeasures. It is worth pointing out that this development does not follow any laws of nature: it is the result of transport policies and human choices.

While Figure 10.1 shows the average modal split of cities and towns, Figure 10.2 shows the modal split for commuter transport in economically highly developed cities on different continents. These

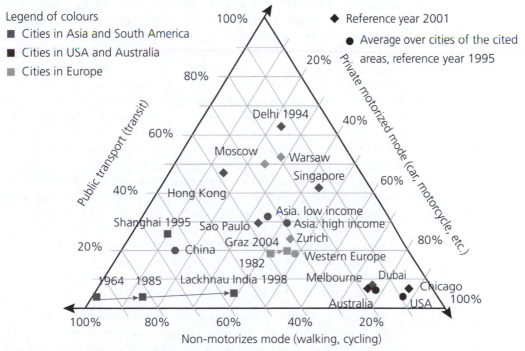

Figure 10.1 Modal split for cities and towns on all continents (shares in trips). (See colour plate 12)
Sources: Kenworthy and Laube (2001); Padam and Singh (2001); Zhou and Sperling (2001)

cities are divided in two groups: into cities with more than 1 million inhabitants (Figure 10.2, right panel) and medium-sized towns with less than 1 million inhabitants (Figure 10.2, left panel). For some selected cities the development of the modal split over time is also indicated, where comparable data are available. When looking at the modal split for these cities and towns it becomes obvious that the range for individual elements of the modal split is quite large, despite similar economic conditions and a high car ownership rate. For example, non-motorized transport ranges from 4 to 50 per cent. If one considers walking as a separate mode, one observes a range from 4 to 21 per cent, while cycling ranges from 0 to 39 per cent. The range for public transport is also high, from 5 to 63 per cent, while private motorized transport ranges from 24 to 80 per cent. Thus, a high share of private motorized transport in towns and cities and an extensive use of cars are not an unavoidable outcome for cities in high-income countries illustrating the importance of urban form and transport policy choices.

The modal split of commuter transport (Figure 10.2) shows also a significant difference compared with all travel purposes (Figure 10.1): the share of non-motorized modes is significantly lower and that of public transport is higher, which is caused mainly by the longer journeys for working trips. There is no significant difference between large (>1 million inhabitants) cities and smaller towns. The development of the commuting modal split over the long term shows a clear trend: motorized modes gain shares at the expenses of non-motorized modes.

A considerable range for the modal split and the associated energy use can thus be observed in towns, cities and urban agglomerations of industrialized countries, and this range is influenced by several factors. These factors can be determined by the inhabitants and decision-makers of a city, provided a strong determination to influence traffic policy and an acceptance of such policy exist. The factors explaining different transport modal splits and their changes include: lifestyles, awareness of environmentally friendly and energy-saving travel behaviour, objectives of urban transport policy and the willingness to implement such policies, spatial structures and settlement density, fuel prices, parking management, provision and operation of transport infrastructures for walking, cycling, public, and car traffic, pricing of the various modes of transport (fees for moving and stationary traffic, public transport prices), as well as priorities accorded to the various modes of transport. The widely held conception that a modern and economically viable city with a high quality of life can only be achieved with car-oriented private transport modes has been demonstrated to be a myth (Newman and Kenworthy 1999) rather than reality.

10.1.2 Modal split and energy use

There exists a specific characteristic amount of energy use for every mode of transport usually measured in terms of litres of fuel or in MJ (megajoule) per passenger kilometre travelled. The specific energy use of each mode is determined by the respective vehicle technology (e.g.

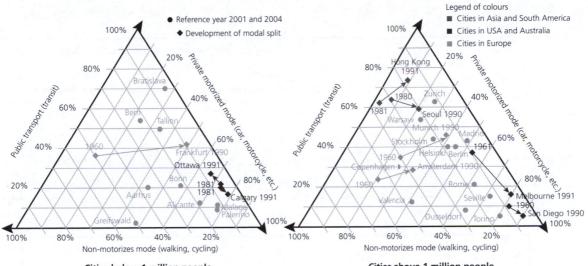

Figure 10.2 Modal split of journeys to work in medium-sized towns with a population below 1 million people (left panel) and in cities with a population above 1 million people (right panel) in high-income economies for reference years 2001 and 2004 and selected time trends since 1960. (See colour plate 13)

Sources: Urban Audit (2009); Vivier (2006); Steingrube and Boerdlein (2009); Wapedia (2009)

size/weight of vehicle, engine efficiency, etc.), the vehicle occupancy rate (passengers per vehicle), as well as traffic conditions (degree of congestion). Overall, the specific energy use is highest for private car transport (Table 10.1). A robust finding in the literature, illustrated in Table 10.1 is that the choice of transport modes (i.e. the modal split) is a more important determinant of overall transport energy use compared to the specific energy use of individual transport technologies. In simpler words: using an (even energy inefficient old) public bus is a more energy efficient mode of transport than using a 'cutting-edge' energy-efficient hybrid private car. Policy-wise, this implies if the objective is to minimize energy use, towns and cities should attempt to achieve a position for their modal-split that favours 'ecomobility', i.e. non-motorized and public transport, rather than incremental improvements in vehicle efficiency or biofuel supply for individual passenger cars, even if the latter may constitute important complementary options for sustainable transport planning in the interim before policy measures in urban form, traffic planning (especially for non-motorized mobility) and for improved public transport systems have a long-term effect.

Table 10.1 Primary energy use per passenger-kilometre travelled for different modes, characteristic ranges for reference year 2005

	Energy use per passenger-kilometre travelled	
	Fuels litre/passenger-kilometre	Energy MJ/passenger-kilometre
Private car traffic	0.050 to 0.075 (100 %)	1.65 to 2.45 (100 %)
Private motor bike	0.028 to 0.038 (55 %)	0.92 to 1.25 (55 %)
Public bus	0.009 to 0.013 (20 %)	0.32 to 0.40 (20 %)
Electric railways and public transport	0.002 to 0.004 (5 %)	0.53 to 0.65 (35 %)

Source: calculation based on Pischinger et al. (1997)

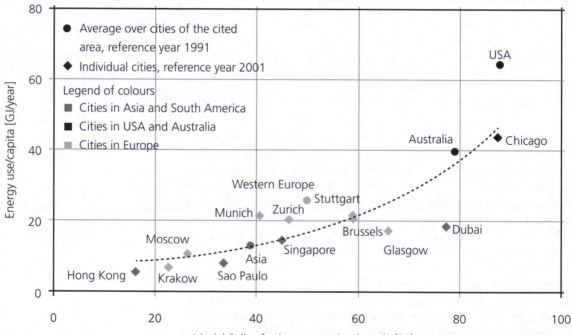

Figure 10.3 Average energy use per capita in transport (average for countries and regions in 1991 and for selected global cities in 2001) versus share of private motorized transport in modal split
Source: Kenworthy et al. (2001, 1999); Vivier (2006)

10.2 Specifics of urban transport systems

Compared to rural and long-distance traffic, urban transport systems have some unique characteristics:

- Towns, cities and urban agglomerations are places of high population and facility density. Their communication needs are high. This leads to a high density of travel demand within limited space. Therefore modes of transport are needed which require little space while offering high performance (Table 10.2). Pedestrians need the smallest amount of space, followed by public transport, cycling, shared taxis and private motorized transport. The space required by the different modes of transport and their capacity depends critically also on occupation rates (load factors). The figures provided in Table 10.2 refer to peak-hour traffic which means full occupation of public transport and about 1.2 persons per car in private motorized transport, characteristic for average load factors during peak commuting times in high income, high car ownership cities.

● As far as noise, emissions, energy use and costs are concerned, non-motorized modes of transport are superior to motorized ones. Due to the limited amount of space and the resulting conflicts between the quality of the environment (noise, emissions, separation effects, etc.) and the use as a living space, non-motorized transport is particularly suitable for urban areas.

Table 10.2 Comparison of characteristic capacities of modes (assuming free-flowing traffic), space required with typical occupancy rates at peak-traffic hours, infrastructure costs in urban settings, and maximum accepted distance for daily trips

	Non-motorized mode		Motorized mode			
	Walking	**Cycling**	**Motor bike Two wheeler**	**Car**	**Shared taxi**	**Public transport**
Capacity of a 3 metre lane [person/h]	3,600 to 4,000	3,600 to 4,000	4,300 to 5,000 (max. 7,200 to 8,500)	2,300 to 3,000[1] (max. 9,500 to 12,000)	5,000 to 9,000	Bus: 8,000 to 16,000 Tram: 18,000 to 24,000 Underground: 30,000 to 60,000
Space required per person [m²/person]	0.7 to 0.8	6.0 to 7.5	13.0 to 15.0	21.0 to 28.0	7.0 to 12.5	Bus: 1.25 to 2.5 Tram: 1.7 to 2.3 Underground: 0.75 to 1.50
Infrastructure investment costs for the space required per person [€/person]	50 to 150	50 to 150	1,500 to 3,000	Urban road: 2,500 to 5,000 Urban motorway: 50,000 to 200,000	1,250 to 2,500	Bus: 200 to 500 Tram: 2,500 to 7,000 Underground: 15,000 to 60,000
Accepted distance for daily trips [km]	1 to 2	5 to 10	10 to 20	Practically unlimited	10 to 20	Practically unlimited
Type of mobility service	Door-to-door service, Temporarily unrestricted service					No door-to-door service, Scheduled service

[1] Average occupancy rate of 1.2 persons/car of peak hour

Sources: Sammer et al. (2009); ÖVG (2009)

- With respect to energy use, emissions, required space and noise, public transport is significantly superior to private motorized transport, if a sufficiently high occupancy rate of public transport modes can be assured. From the transport user's point of view public transport has certain disadvantages compared to private motorized transport which can only partly be compensated for. Such disadvantages are the limited spatial and temporal availability due to a scheduled service with specific stops (Table 10.2). To make public transport attractive, a dense public transport network and a high service frequency with short intervals are necessary. One should attempt to achieve headway of service of fewer than 10 minutes, so that passengers are able to use public transport spontaneously. To make access and egress attractive, the catchment area of a public transport stop should not span more than 300 to 500 metres. This can only be guaranteed if the densities of development of residential areas in cities (including traffic areas) are considerably higher than roughly 100 persons per hectare (10,000 persons per km²) and densities of development within the whole urban area are higher than 50 persons per hectare (5,000 persons per km²). Because of the large amount of space required for road areas and the low-density housing types (e.g. single-family dwellings) that preordain private car use, car-dependent urban settings have typically population densities far below these critical threshold levels (Figure 10.4). An increasing share of private motorized transport modes causes a progressive increase in urban space and a progressive decrease in urban density. Since their density of development is unsuitable for offering attractive and economically viable public transport systems, car-dependent cities are 'locked into' high energy use (and emissions).

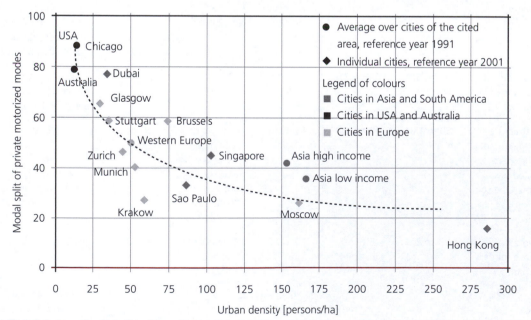

Figure 10.4 Relation of urban density and share of private motorized transport modes (calculated from total mobility, including non-motorized modes) for individual cities and regional average cities
Sources: Kenworthy et al. (2001, 1999); Vivier (2006)

- In terms of spatial planning, clear objectives with respect to sustainable urban transport systems can be derived: stores for items of daily use need to be in locations which can be reached by non-motorized transport modes. The catchment area of such stores needs to be characterized by a sufficient density of use and compact settlement structures to make these stores economically viable. Thus, a concentration of stores for daily needs in shopping centres in car-oriented locations at the city perimeter should be avoided at all cost. The growth of out-center shopping and other facilities leads to a negative feedback loop that further reinforces automobile dependence: fewer neighbourhood stores, rising demand for shopping centres, increasing use of cars, and yet fewer customers for local stores (Figure 10.5). A similar negative feedback loop can be observed in the cause–effect chain from road infrastructure development, generation of more car traffic, environmental damages, lower urban quality of life, migration to suburbs, to yet further generation of more car traffic. It needs to be stated that contrary to popular conception any additional supply of urban road infrastructures (a frequent popular response to congestion by dense car traffic) inevitably increases demand as well, entailing more use of cars, ultimately simply shifting the point of congestion to higher levels of road and car traffic densities (Sammer et al. 2009).
- Since trip lengths in local traffic are considerably shorter than those in long-distance traffic, high transport capacities at more moderate speed but with high travelling comfort are most important for short-distance, i.e. urban, traffic.
- From an economic point of view, private motorized transport and public transport do not compete on a level playing field since the external costs of the two modes of transport differ considerably.

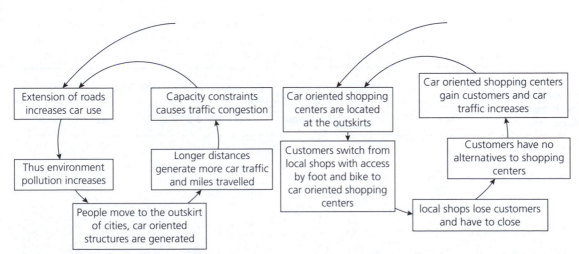

Figure 10.5 Dynamic negative feedback effects between urban sprawl, car traffic, the quality of the environment, and viability of neighbourhood stores in urban areas
Source: Sammer et al. (2009)

Table 10.3 offers an illustrative comparison of the external costs of road traffic and rail transport in Austria and Germany, where available data and estimation methods allow for one of the few 'apples-to-apples' comparisons possible in the extremely heterogeneous external-costs literature. Compared to rail transport, road traffic generally causes five to six times higher external costs. The overall results for Austria and Germany are quite similar but there are significant differences as far as some of the cost components are concerned, mostly caused by different assumptions with respect to external cost rates per externality category. (Thus even in this case a fully consistent comparison of external costs is not possible: in the relevant literature – e.g. UNITE (2003); Maibach et al. (2008) – external-cost estimates from transport vary considerably, suggesting the difficulty of arriving at widely agreed consensus values useful for policy makers.) The external costs caused by private car uses have to be paid for by other people. As far as the costs of noise, emissions and accidents are concerned, the people who finance the health-care system have to pay for them, irrespective of whether they use a car frequently, infrequently or not at all. The resulting costs of greenhouse gas emissions caused by fossil-fuel use in cars have to be paid by future generations that will have to face the consequences of climate change. From the twin perspectives of sustainable transport systems as well as from an economic perspective of a fair, level-playing field, competition among different transport modes externalization of costs needs to be overcome via internalization. Polluters should bear the resulting external costs by making them pay suitable fees or taxes, which in reality would translate into an economic effect equivalent to a fuel price increase of at least a factor of 2–3, which does not look politically feasible at present. This suggests a more incremental policy strategy of a gradual phasing in of external costs.

Table 10.3 Estimates of external costs of road traffic and rail transport for passengers and freight in Austria and Germany (2005). For comparison: average total car operating costs (incl. taxes) per person/kilometre (paid by the driver) are about 24 €-cent per person/kilometre, the share of fuel/cost is 6.5 €-cent of that (2005)

External costs in €-cent per person-kilometre and ton-kilometre	Austria		Germany	
	Road traffic	Rail transport	Road traffic	Rail transport
Environment costs (caused by greenhouse and exhaust gas emissions, noise)	8.6	0.6	3.3	?
Accidents	3.1	1.5	5.1	?
External costs total	11.7	2.1	9.4	2.0

Sources: Sammer (2009a); Pischinger et al. (1997)

10.3 Cost-effectiveness of measures to reduce transport energy use

Effective policy intervention in transport systems needs to know how much, by what measure, and at what costs fossil-fuel use and associated environmental externalities can be reduced. In addition political acceptance and the distributional effects of policy measures are also of crucial importance for their implementation feasibility.

The following section has the objective of illustrating how the concept of cost-effectiveness can be applied to a wide range of transport policy options. While not necessarily generalizable at a wider geographical scale, the illustrative example below for Austria demonstrates that methods and data are in principle available to move transport policy choices to a more rational ground. Pischinger et al. (1997) and Sammer (2008) have analysed in detail twenty-five different policy measures to reduce fossil-fuel use in transport systems. The measures analysed include travel demand management (TDM), land use and regional planning, road and railway infrastructure improvement, as well as improvements in vehicle technology (Table 10.4). The macroeconomic costs include the operating and investment costs of the measures, the vehicle operating costs, time and environmental costs, accident costs (i.e. external costs) as well as the consumer surplus of the induced travel demand.[1] The cost-effectiveness of policy measures are presented as the ratio of the macroeconomic cost (cost minus benefit) per reduced ton of fossil fuel, with the absolute potential for fuel use reductions shown as well (Figure 10.6)

The analysis of the cost-effectiveness and the potential of reducing fossil-fuel use produced the following results: TDM measures are generally highly cost-effective compared to other types of measures. But their potential to reduce fossil energy use varies depending on the measure considered. Measures which increase travel costs significantly (e.g. fuel surcharges or road pricing), have a considerable reduction effect (but do not seem easily acceptable to the public). Conversely, area-wide promotion of bicycle use, voluntary programmes to change travel behaviour as well as land use and regional planning measures are rather cost-effective but have only a small short-term absolute reduction potential. All above-mentioned policy measures have a negative cost-effectiveness ratio (i.e. reduce fuel use as well as costs, i.e. yield a net macroeconomic benefit, incl. external costs), which is not the case for the type of policy measures addressed next. Investments in road and railway infrastructure or public transport are hardly cost-effective and their potential to reduce the fossil energy use in transport is also rather low. Such measures can only be justified to remove bottlenecks and traffic congestion in order to improve the accessibility for private motorized vehicles, which however due to the additional car usage induced defies the overall policy objective.

Table 10.4 Overview of bundles of policy measures analysed in an illustrative cost-effectiveness study on the reduction of fossil energy use of motorized transport in Austria

No.	Bundles of measures: *Travel demand management*
1	More effective enforcement of existing speed limits
2	Lower speed limits (30/50/80/100 km/h)
7	Extension of pay parking management in cities
8a	Eco-Bonus with doubled fuel price and distribution of additional revenue to the population
8b	Environmental fuel charge doubling fuel price used for financing of environmental protection measures
8c	Area-wide road pricing with the effect of doubled fuel prices
9	Extra charge for air traffic take-offs and landings (€ 17 per seat)
10	Area-wide promotion of bicycle use (hard and soft measures)
15	Improvement of freight logistics (setting up an 'inter-modal logistics association')
16	Intensification and extension of ITS (Intelligent Transport System)
22	Voluntary programme to change travel behaviour
No.	**Bundles of measures: *Land-use and regional planning***
4	Avoidance of urban sprawl (site development levy of real estate for local transport infrastructure and its operation, e.g. city toll)
5	Residential building subsidy, dependent on distance to stops of public transport
6	Subsidy for land purchase, dependent on distance to stops of public transport
No.	**Bundles of measures: *Road and railway infrastructures***
10	Extension of pedestrian districts with corresponding restrictions to car traffic
12	Extension of inter-modal freight transport (terminals, network)
13	Extension of passenger transport of railways (regular headways)
14	Extension of urban public transport (tram, bus, underground, etc.)
No.	**Bundles of measures: *Technology of motor vehicles***
17a	Reduction of the fuel consumption of cars (ECE 1/3-mix)
17b	Reduction of the fuel consumption of cars (obligatory limit of 3 litres/100 km)
18a	Zero-emission vehicles (electric cars for the public sector)
18b	Zero-emission vehicles (electric cars as 3% of new registration)
19	Extended use of trolleybuses (replacement of 20% of urban buses)
20	Bio-fuel (Austrian capacity of 210.000 tons/year)
21	Intensification of vehicle inspections

Source: Sammer (2008)

The cost-effectiveness of measures to improve vehicle and engine technology included in the Austrian case study is comparatively low and their reduction potential is also limited, except in the case of a mandatory maximum fuel use standard (3 litres/100 km). Compared to this, stricter speed limits on motorways as well as main and secondary roads have a very good reduction potential and their cost-effectiveness is also comparatively high. It is interesting to look at the impact of telematics-supported traffic guidance systems. They have little impact upon the reduction of fossil fuel use and their cost-effectiveness does also not compare favourably.

The results – while not necessarily generalizable in all their details – suggest that generally 'soft' transport planning measures have generally a very high reduction potential and also are highly cost-effective, but they are politically the most difficult to sell. Fuel use for transport can effectively be reduced by well-coordinated 'bundles' of measures as opposed to fragmented individual policy initiatives. Such comprehensive policy 'bundles' should include financial charges and other 'soft' measures influencing travel demand, with the overall objective of internalization of external transport costs. Since the political acceptance of any noticeable financial surcharge seems currently quite low, there is an urgent need to raise public awareness first. It must be stated that the result of the ranking in Figure 10.6 can change totally when leaving out the external costs.

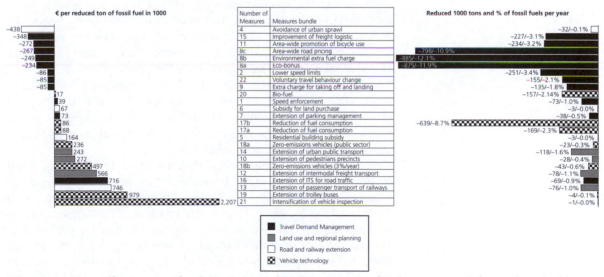

Figure 10.6 Cost-effectiveness of policy measures to reduce transport fossil energy use (left panel, ranked by cost-effectiveness) and absolute and relative reduction potentials (right panel) for Austria, by measure (2005). The cost-effectiveness is presented as the ratio of the macroeconomic cost (cost minus benefit, including external costs) per reduced ton of fossil-fuel use. For policy measures listed, see Table 10.4.
Source: Sammer (2008)

10.4 Strategies and frameworks for effective policy measures in cities

The following section summarizes the strategies, basic conditions and measures to consider in policies for effective reduction of urban transport energy use. They do not address issues of vehicle technology and alternative fuels and the basic conditions for their use.

10.4.1 Strategies to take eight relevant decisions to encourage energy-efficient travel behaviour

These strategies are based on a systematic approach which takes into account the interaction of human behaviour, travel demand, transport infrastructure supply, vehicle technologies as well as the environment and financing.

The starting point is human decision-making which has a significant impact on urban transport systems and their energy use. The following eight types of decisions by transport users have a crucial impact on urban transport systems and their effect upon the ecology, economy and society. One needs to bear in mind that these decisions are usually taken in some sequence, that feedbacks takes place, and that every decision taken has an impact upon the next decision.

1. Selection of the place of residence and place of work. The closer these two locations, the better equipped with local stores and other facilities and the more ecomobility can be used, the less frequent motorized private transport, the lower transport energy needs. To make this possible, compact settlement structures are needed; a minimum settlement density of well above 50 inhabitants per hectare (5,000 inhabitants per km2) should be the policy objective.
2. Selection of the availability of vehicles in a household. Ownership, availability and types of vehicles (bicycle, motorbike, cars with different drive technologies and fuel use, season tickets for public transport, membership in a car-sharing organization, etc.) have a significant impact on travel behaviour.
3. Decision: whether some physical distance needs to be covered (trip) or whether an activity can be done at the place of residence or handled with the help of telecommunication technologies; good facilities at the location, for example broadband ITC infrastructures and/or a garden can help to reduce the need to travel any distances.
4. Decision about the time of travel. Flexibility regarding the time of travel helps to save time, money and energy and to use resources in an environmentally friendly way, because traffic jams and an overcrowding of public transport can be avoided. This implies that wherever possible over-'synchronization' of social activities should be avoided, e.g. through 'stacked' timings of school and workplace starting and ending times.
5. Choice of destination. Good retail, work and school infrastructures close to the place of residence and compact settlement structures

help to avoid long commuting and shopping trips and the need to use motorized transport.

6. Modal choice. Compact developments with good public (e.g. schools) and private (e.g. retail) infrastructures at the place of origin and destination as well as good connections between the two with an attractive ecomobility offer (walking, cycling and public transport) or a suitable offer of intermodal transport (bike-and-ride, park-and-ride, park-and-drive, park-and-bike, etc.) help to avoid unnecessary car trips. The quality of door-to-door connections is crucial for modal choices. To offer good quality, an attractive area-wide transport network for the non-motorized and public transport is necessary.

7. Route choice from origin to destination. The concept of environmental zones, i.e. protected zones for residential areas, etc., in connection with a hierarchical road network for motorized private transport and transport of goods which has been structured according to the principles of traffic calming improves the quality of the urban environment. It helps to avoid through traffic of motor vehicles in protected areas and supports a high quality of housing and traffic safety.

8. Reflections about and revisions of past decisions regarding travel behaviour in regular intervals. Were they appropriate? GPS technology in connection with new traffic information technologies will soon make it possible to check automatically whether decisions about traffic behaviour are appropriate for the specified objectives (time requirements, environmental friendliness, cost of transport, etc.) or whether alternatives are preferable. This kind of individual mobility information system can lead to more sensitivity and awareness of transport-related decisions.

10.4.2 Principles and frameworks to reduce the fossil energy use in transport

- There is no single, 'silver bullet' measure to reduce transport energy use effectively; a whole range of well-coordinated measures is needed. Traditional 'supply-side' measures which focus on infrastructure supply are both insufficient to solve urban transport problems and risk significant consumer 'take-back' effects (i.e. induce additional car mobility rather than reducing congestion).

- A new fact-based systemic decision-making culture is needed for urban transport policy. Policy-makers need to be prepared to suggest and to implement unpopular but necessary measures (e.g. internalization of external costs). It is essential that mere reactive and adaptive planning by individual measures without consideration of systemic effects and feedback (e.g. demand responses) such as the often ill-considered simple extension of road networks in urban areas (so-called 'adaptive planning') is replaced by some systemic and goal-oriented planning procedure.

- As long as the true cost of various modes of transport, particularly motorized private transport and public transport, are not transparent to the users due to the lack of internalization of external costs, unfair competition between transport modes is perpetuated. The 'push-and-pull principle' by restricting car traffic while at the same time enhancing ecomobillity modes can help to compensate for lack of internalization of external costs.
- Cost-effectiveness criteria should be used to select policy measures to reduce fossil-fuel use and to make effective use of limited available financial resources.
- An effective programme to reduce urban transport energy use can only work as an integrated concept which takes all modes of transport into account. For the concept to be successful, regularly supervised and adapted quantitative objectives for the reduction of energy use are essential.
- To safeguard political acceptance, development and implementation of the programme need to be supported by some suitable measures to shape stakeholders' ideas and guarantee their participation and engagement (Kelly et al. 2004).

10.4.3 Effective measures to reduce fossil energy use in urban transport

- Spatial planning. The creation of compact settlement structures with a sufficient settlement density well above 50 people per hectare is essential to offer attractive non-motorized and public transport mode options. To this end, regulatory measures alone are insufficient; some market-based policy instruments are needed in addition, such as charging the development cost to users in areas of lower density.
- Integrated planning concepts to save energy in transport. Since no single 'silver bullet' measure exists to reduce urban transport energy use, successful programmes including whole bundles of measures need to be planned at national, regional and local levels, and implemented in a continuous process. Coordination and harmonization of transport policy across all levels is necessary. A thorough analysis of the real situation must be the basis for an efficient implementation programme to develop the best possible goal-oriented measures for every specific location.
- Internalization of external costs. In principle internalization of external costs is essential for all modes of transport to make the polluters pay for the cost they cause. This is an excellent market-based instrument to reduce fossil-fuel use. This can be achieved by fuel price surcharges, but also by parking fees and road pricing. The latter policy option is most effective if linked to the distance travelled and the environmental damage caused. Reduced fees for car pools or a variable and dynamic kind of road pricing depending on utilization rates and congestion status (Supernak 2005) create additional incentives to save energy. If road tolls are only used in urban areas or

certain areas within cities, there is a high risk that undesired side-effects might occur, such as shifting routes to side streets, moving of stores and other facilities to the outskirts where no fees are levied and in the long run even a relocation of companies to areas outside cities (Sammer 2009b) which would lead to undesirable urban sprawl.

- Parking management schemes in towns and cities. Measures to limit parking in cities and towns which cover all densely populated areas are highly effective. Such measures include parking fees on public streets; they are particularly effective if they are graded depending on environmental friendliness or energy use of cars. Since there is considerably more private than public parking space in many towns and cities, it is recommended to include also large private car parks (e.g. of industrial enterprises) into the parking-fee scheme and combine this with levies on parking spaces for their operators (Sammer et al. 2007). In areas with a well-developed public transport system land-use and zoning planning should fix an upper limit for available parking slots instead of (the customary) minimum number of mandated parking spaces (Sammer et al. 2005).

- Public transport. An attractive public transport offer must provide convenient door-to-door connections. This means that access and egress distances of stops and distances between stops and exchange points as well as the time spent in vehicles need to be considered as a whole. The transport user wants a qualitatively high door-to-door mobility service. He does not care which operator is responsible for different parts of the system. Therefore the good cooperation of all public transport operators, an integrated ticketing and timetable system and some overall responsibility for the integrated public transport system are essential for an attractive system. There is an obvious need to improve public transport in regard to intermodality, more efficient and rationalized operations, intermodal connections, information and marketing. To compensate for some disadvantages, public transport should be given priority over motorized private transport by the provision of high-occupancy-vehicle lanes (HOV lanes) and priority treatment at crossings with the help of suitable traffic lights. To make public transport more attractive, good synergy effects can be achieved by linking the system to non-motorized traffic (bike-and-ride, etc.) and by combining the measures already mentioned with restrictions of motorized private transport (push-and-pull-strategy).

- When choosing types of public transport a careful check of costs and effectiveness is needed, since investment and operating costs of various systems differ considerably: in general, buses tend to be considerably cheaper than trams or suburban railways which require tracks, and underground track-bound systems are more expensive than above-ground systems. In general, renovating existing transport systems is most cost-effective, for example by eliminating existing obstructions by car traffic or by giving public transport priority over motorized private transport with the help of HOV lanes and traffic lights.

- Non-motorized traffic (walking and cycling). Measures to encourage and support non-motorized traffic are a highly efficient way to save energy, particularly if they are combined with support for public transport and restrictions for car traffic. Suitable measures are the extension of an area-wide network of walkways and cycling routes and making them more attractive to use, more places for leaving bicycles in public and private places, more information, marketing and measures to shape people's ideas, bicycle renting, permission to transport bicycles in public transport systems, company bicycles, etc. (Meschik 2008).

- Access restrictions for motorized private transport. Environment-oriented access restrictions for cars in city centres are highly effective measures to save fossil fuel, for example the temporally limited or unlimited prohibition of access for certain types of vehicles as in environmental zones to keep the air clean (Umweltzone Berlin 2009), e.g. through access contingents for cars with combustion engines or through an area-wide traffic light system. The latter means that access is restricted by traffic lights in such a way that only so many cars are allowed to pass, thus avoiding potential congestion at subsequent crossings regulated by traffic lights. Outside areas with such traffic light management additional space needs to be provided for car traffic, as well as HOV lanes and park-and-ride facilities.

- Roadworks. If the goal is a reduction of fossil-fuel use, extending the existing road infrastructure, as is frequently done in urban areas, cannot be recommended. Every extension of existing capacities induces an increase in motorized private transport, and further substitution away from other, more environmentally friendly transport modes.

- Highway corridor management. This concept has the objective of optimizing the traffic flow by suitable measures such as HOV lanes, high-occupancy-and-toll lanes (HOT lanes) by variable pricing, ramp-metering, information about suitable times, etc. Such management systems should help to avoid congestions and time loss due to congestion. It is doubtful whether they help to reduce energy use significantly in the long term. At present some so-called 'Integrated Corridor Management Projects' (RITA 2009) are conducted in the USA; they try to include public transport, for example by using emergency lanes as bus lanes.

- Mobility management at the enterprise level. This is a highly efficient tool. It includes all kinds of incentives to make individuals do without cars. Suitable measures are the support of car pools by providing incentives for their use (guaranteed home transport if the usual transport is missed, preferential parking slots, financial incentives), so-called job tickets for public transport (employers subsidize season tickets or provide them free of charge), repair and shower facilities for cyclists, etc. The effect can be increased by providing HOV lanes or by making large companies develop mobility management plans (ICARO 1999). In-company mobility management leads to a three-way win-win-win situation since employers, employees and the general public all benefit. For mobility management at the company level to be successful, a permanent management process must be professionally run.

- Voluntary programmes to change travel behaviour. Such programmes are quite efficient at raising awareness (travelsmart Brisbane, 2009) of the impacts of energy-saving traffic behaviour. They have the potential of reducing the use of fossil fuels of the target population in urban areas by 5 to 10 per cent. Suitable ecomobility alternatives must be available for such programmes to work. Currently some attempts are made to combine such programmes with energy-saving measures in private households (Brög and Ker 2009; DIALOG 2010).

- Electric mobility as a future chance to reduce fossil energy consumption in transport. At present, electric mobility is seen as one of the main important energy solutions in transport, but the technological breakthrough is yet to come. The missing links can be summarized as (Link et al. 2011a, b):

  o narrow range of electric vehicles due to inadequate battery technology; even if the average distance travel per day is below 100 km, for the user the definite expected range is dependent on the user's potential need of range over the lifetime of his car and this range is close to his experience with his conventional car;

  o total life-cycle costs of an electric vehicle are about twice the costs of a comparable conventional car; it can be expected that after the technological breakthrough the investment costs of an electric car will decrease substantially;

  o potential to produce enough electric power from renewable and regenerative energy sources in order to run all electric cars.

So, electric mobility is still a challenge for the future, not a short-term solution.

10.5 Conclusions

The modal split of cities is one of the key determinants of urban transport energy use and also a good indicator of progress towards improved sustainability of urban transport systems. With an increase in private motorized transport, comes a progressive increase in energy use. A wide range of modal split and thus energy use patterns can be observed across different urban settings. This variation in urban mobility is not ex ante given, but rather results from deliberate choices of individuals and of decision makers. The factors determining urban transport choices can be changed, provided there exists a strong determination for sustainable traffic policy and a corresponding wide public acceptance of the overall goals of such a policy. This wider acceptance of the overarching goals is also the condition to implement some individual measures (e.g. traffic calming, parking fees, etc.) that often face public opposition.

An urban transport system has specific characteristics. As far as noise, emissions, energy use, and economic costs are concerned, non-motorized modes of transport are superior to motorized ones, and public transport is significantly superior to private motorized transport, provided a sufficiently high occupancy rate in public transport can be achieved. To make public transport attractive, a dense public transport network and a high service frequency with short intervals are necessary. This can only be guaranteed if the densities of development of residential areas in cities are considerably higher than roughly 100 persons per hectare and densities of development within the whole urban area are higher than 50 persons per hectare. Because of the large amount of space required for road areas and low-density housing types (e.g. single-family dwellings) that preordain private car uses, car-dependent urban settings have typically population densities far below these critical threshold levels.

A remarkable number of suitable strategies and measures exist to reduce urban transport energy use. A very effective strategy is the promotion of high urban densities in combination with active promotion of a high quality supply of non-motorized and public transport options combined with a restrictive car policy. Soft measures have generally a very high potential and high cost-effectiveness to reduce energy use but they are politically the most difficult to sell. Transport fuel use can be most effectively reduced through well-coordinated bundles of policies and measures. Key is the internalization of external costs of motorized private transport combined with the provision of a high-quality alternative offer of public and non-motorized transport. To be able to internalize external costs, the political acceptance of this unpopular measure must be increased. In the medium term, the trend towards urban sprawl must be stopped to generate a sufficiently high settlement density to be able to reduce automobile dependence.

Note

1   Induced travel demand means new generated trips or increased travel distance by improved accessibility as the consequence of improved transport infrastructure. On the one hand the induced travel demand causes a benefit known as consumer surplus and on the other hand it causes negative environmental impacts.

11

Urban energy systems planning, design and implementation

James E. Keirstead and **Nilay Shah**

11.1 Introduction

The energy system elements and networks of a city reflect myriads of 'local optimizations'. The networks have thus evolved over time but seldom exploit the opportunities for broader optimizations with other networks or urban form. They are consequently not usually resource efficient when viewed in aggregate.

Fortunately systems and integration methods are now becoming available which promise the potential for reductions in direct primary energy consumption of 20–50 per cent without other significant physical impacts except the advantages of reductions in externalities of energy use such as air pollution. A popular application of such techniques is cases where the energy service demands of a city are known and the design problem is to select an optimal mix of technologies and fuels to meet these demands. Examples include an analysis of renewable energy supply options for a UK eco-town (Weber and Shah 2011), an examination of investment staging for urban energy policy (Bruckner et al. 2003), an assessment of the cost-effective emissions reduction potential of an urban area (Brownsword et al. 2005), and an investigation of building, district and city-scale options for transforming the Japanese commercial buildings sector (Yamaguchi et al. 2007).

Some work has also been done on optimizing aspects of urban form, although these studies typically focus on transport energy consumption only. This approach has been applied at the master-planning stage, as with 'sketch' modelling frameworks (such as Lin and Feng 2003), or to assess the inefficiencies of existing transport patterns, as in the excess commuting literature (e.g. Ma and Banister 2006).

11.2 Modelling urban systems: the SynCity model

Given the complexity of a city's energy and transport systems, it is not surprising that detailed holistic analysis of the interplay between urban form, a city's built environment, energy demand characteristics, and its transport and energy systems have not been attempted to date.

Bottom-up assessments of energy-efficiency improvement potentials in different sectors have been developed for many cities in support of policy choices, but interactions both in terms of potential synergies or tradeoffs cannot be explored by such compartmentalized approaches.

New computational modelling frameworks and access to new data sources promise to overcome these barriers. The relationship between key parameters such as population density and energy may be obscured in the real world by differences in other factors such as wealth and income. In order to explore these interactions under comparable *ceteris paribus* conditions, as apart of this assessment, illustrative model simulations were commissioned with one such modelling framework: SynCity (see Box 11.1), the results of which are reported below.

In these examples, the synthetic city ('SynCity') is an urban settlement for 20,000 people in a service-oriented local economy, in a moderate climate with gas (and oil) as the primary fuel. Five SynCities have been explored with the optimizer, varying in four key respects:

1. **Building fabric**   The quality of the built environment is represented by the UK Standard Assessment Procedure rating (SAP, BRE 2009). A SAP rating of 50 represents a medium level of efficiency in space and water heating, as well as lighting, and corresponds approximately to the current standard of London's housing stock (GLA 2004). For a high efficiency design, a SAP value of 100 was chosen which corresponds to the Passivhaus standard of 15 kWh/m² year for space heating. Higher SAP values are possible by including onsite microgeneration but these are not considered at this stage of the analysis.
2. **Density**   The layout model can choose from a range of residential housing types, each with a different density and floor area. Three density archetypes are used here. The UK type describes densities of 20, 35 and 65 dwellings per hectare and medium-sized dwellings (60–200 m²). The Japan type represents higher densities of 50, 75 and 100 dwellings per hectare and smaller dwellings (30–100 m²). Finally, the US type represents the low-density sprawl found in many North American cities with housing densities of 5, 10 and 20 dwellings per hectare and associated larger floor areas of (200–300 m²).
3. **Electricity power**   The energy system is modelled as either a (UK) 'business-as-usual' system, with heating provided at the household scale by gas boilers or electric heaters, or a CHP-based system with gas-fired combined heat and power systems at three different sizes (1, 3 and 6 MW thermal) and an associated district heat network.
4. **Layout**   In cases where the layout is not optimized, a simple mononuclear city has been assumed. In the optimized cases, the layout model can choose the position of dwellings and activities subject to key constraints, such as ensuring that all required services are provided and there is adequate housing for the entire population.

Box 11.1 The SynCity toolkit

SynCity (Synthetic City) is a software platform for the integrated assessment and optimization of urban energy systems, developed at Imperial College London and supported by funding from BP. The goal of the toolkit is to bring together state-of-the-art optimization and simulation models so that urban energy use at different stages of a city's design can be examined within a single platform.

There are three layered models within the system used here:

1. a **layout** model, which determines the optimal configuration of buildings, service provision and transportation networks;
2. an **agent–activity** model, which simulates the activities of heterogeneous agents acting within a specified urban layout in order to determine temporal and spatial patterns of resource demand; and
3. a **resource technology network** (RTN) model, which determines the optimal configuration of energy conversion technologies and supply networks.

The **layout model** is a mixed integer linear programming model which seeks to satisfy urban demands for housing and activity provision, while minimizing energy demand from buildings and transport. Users specify average visit rates for each activity type and the model will determine the optimal location for housing, commercial buildings and activities, and transport networks.

The **agent–activity model** is a simulation model designed to estimate the resource demands of a population living with a particular city layout. Briefly, the model operates as follows. First, the model creates a synthetic population of individual agents with random characteristics such as gender and education. Agents are grouped into household ensembles and assigned to jobs and dwellings. The model then loops over sixteen indicative time periods representing two seasons (summer, winter); two day types (weekday, weekend); and four time intervals during the day. For each interval, a probabilistic four-step transport model is used whereby citizens select an activity, an activity provider, a transport mode and a travel route. The agents then move around the city, performing their planned activities, resulting in spatially and temporally explicit demands for different end-use energy resources, such as electricity or heat.

The **RTN model** is also a mixed integer linear programming model. Its aim is to determine the optimal configuration of energy supply technologies in order to meet a given pattern of demand. The objective is to minimize the total cost of the energy supply system, comprising the annualized cost of capital equipment (e.g. boilers, turbines and distribution networks) and the annual cost of imported resources necessary to operate the system (e.g. supplies of gas and electricity). Users specify the full suite of possible technologies at the outset and the model will return the lowest cost system configuration.

For further details on the methodology, see Keirstead et al. (2009).

These attributes were combined to create representative scenarios. At one extreme the optimization is constrained as a low-density city, fed from a power grid, with modest building fabric energy performance. This city is taken as characteristic of one that has evolved in an economy where resources are relatively inexpensive, such as the USA. The optimizer is left to choose the location of fixed infrastructure to minimize transport costs. The city at the other extreme is optimized with only a constraint on the lower bound on population density. It is comparable with an economy that is resource efficient, such as Japan's. The location of housing and commerce, and the choice of whether to use embedded generation of power, is left to the optimizer. Three intermediate cases are considered based on an intermediate density and imposed mononuclear layout (e.g. United Kingdom).

The increased density of the compact urban layout case means that individual dwellings are smaller and with less external wall area per dwelling, resulting in reduced heating demands (also one-third savings as in the case of the high-standard fabric implemented for all buildings but at lower densities). In other words, urban density and form and efficient building design both yield comparable energy demand reductions in the simulations, highlighting the importance of considering both policy options simultaneously as otherwise there is a risk that the efficiency improvements of better insulated building structures can easily be compensated for by a shift towards less compact settlement patterns. Conversely the construction of large houses in a low-density sparse layout increases these heat losses (one-third increase in primary energy) and also substantially increases transport energy use. This 'suburbanization' scenario, a worst-case scenario in the simulations would result in an almost threefold increase in energy use compared to the optimized solution. In other words, building a Passivhaus standard single-family home in a low-density (sub-)urban area would not lower energy use substantially compared to remaining in a much less efficient home but that is located in a more compact urban setting (e.g. a nineteenth-century townhouse located close to education, leisure and shopping facilities) with its corresponding lower individual transport needs. This is clearly shown by the alternative simulated urban layouts of the model runs (Figure 11.1).

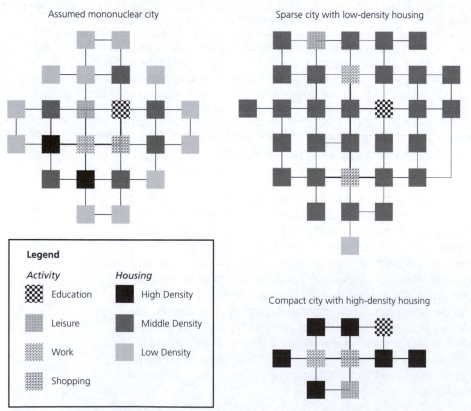

Figure 11.1 Urban layouts (left to right): the assumed mononuclear city, a compact city with high-density housing, a sparse city with low-density housing. In each figure, the coloured cells represent activity provision: green for leisure, blue for work, pink for shopping, and yellow for education. The grey cells represent housing at different densities, with the labels indicating the density in dwellings per hectare. The black lines connecting the cells indicate road connections and indicative traffic flows. (See color plate 14)

It is important to note that only passenger transport flows have been analysed here. In reality, urban transport is a mix of passenger and freight journeys; in London it is estimated that 23 per cent of CO_2 emissions from transport are attributable to freight (including services) and this traffic also creates significant local air pollution (Tfl 2007). The two modes are closely linked. New opportunities in e-commerce, for example online shopping and delivery services, can help to reduce passenger transport flows but may result in increases in freight travel. A significant challenge for future research is therefore to explore these interactions and examine the opportunities for meeting the service requirements of citizens more efficiently.

The results for each city are presented in Table 11.1 and in Figure 11.1. Numerical values are indexed to the annual primary energy consumption of the sparse city design (144 GJ/capita in the simulation). 'Upstream' energy is energy expended at power stations to supply grid electricity to the city. 'Delivered' energy is energy delivered to stationary infrastructure, including combined heat and power plants, and the final end-users (i.e., a combination of final and secondary energy). The total delivered energy, transport and 'upstream' energy corresponds to the customary reporting of primary energy use.

The model results generally support the interpretation of comparative city data presented earlier. First, 'upstream' energy loss in power generation represents 20 per cent of the primary energy consumed where the grid is used. To ignore this contribution and focus only on delivered energy misses important upstream implications of energy choices for power. Second, a SynCity with low resource efficiency is likely to consume about twice as much primary energy as a city designed for high resource efficiency. Both transport and primary energy for heating and power are reduced by about the same proportion. This, in part, reflects that low-density cities not only require higher speeds of travel over longer distances, but that buildings tend to occupy larger areas with consequently more exposed surface for the same standard of construction. The effects of urban planning and differences in fabric standards are comparable and should be seen together with upstream consequences.

Table 11.1 Primary energy use of five alternative urban designs for a town of 20,000 inhabitants. Results are indexed with sparse city =100

Type	Building fabric	Density limit	Electrical Power	Layout	Transport	Delivered	Upstream	Total
Sparse	Medium	US	Grid	Optimized	26	57	17	100
Distributed generation	Medium	UK	Chp	Mononucl	17	52	0	69
Efficient buildings	High	UK	Grid	Mononucl	17	28	12	57
Compact layout	Medium	Japan	Grid	Optimized	12	28	10	50
Optimized	High	Japan	Optimized	Optimized	12	31	0	43

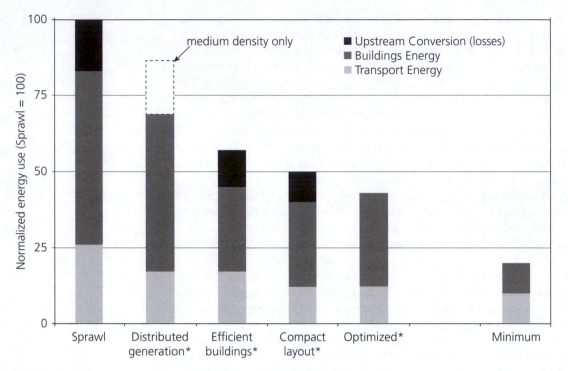

*Medium Density

Figure 11.2 Energy use for five alternative urban designs by major energy level and type normalized to Sprawl city (index = 100) energy use (144 GJ/capita). See Table 11.1 for definitions of the five simulations. The 'Minimum' urban energy use estimates refers to implementation of the most efficient building designs and transport options available and which could not be considered in the scenario simulations

To address these three policy fields simultaneously is also of prime importance for the economic viability of cogeneration and district heating systems. Energy-efficient single-family homes located in low density suburban settings are unlikely to yield the head load densities required for installing capital intensive cogeneration systems combined with local district heating grids (not to mention large centralized cogeneration/distribution systems, although these may allow room for other technologies such as PV or ground source heat pumps). To test the consequences these city models have been rerun to minimize whole life costs. The layout model (see Box 11.1) optimizes transport costs, but it is instructive to see what the resource technology model makes of the stationary energy service costs. The results in terms of total life-cycle costs are summarized in Figure 11.2 using UK electricity and natural gas costs as examples. The discount rate used in the simulation was 6 per cent.

The results encouragingly follow a similar pattern to the primary energy minimization, in part because capital costs are still a small part of the annuitized energy costs. However, minimizing just capital cost biases the outcomes away from the minimum energy solution. This emphasizes the importance of proper finance, pricing systems and business models in bringing about optimal solutions for utilities and customers combined.

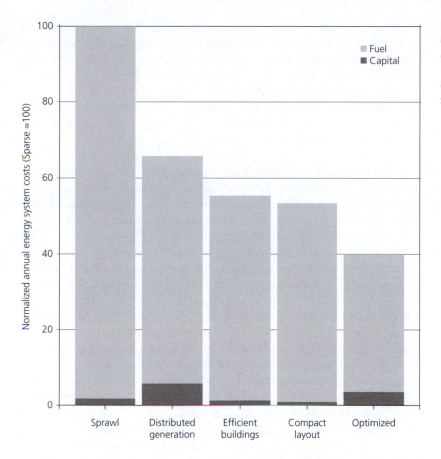

Figure 11.3 Total life-cycle costs (capital plus fuel) of the five city designs indexed to Sprawl city = 100 (see Table 11.1 for definitions of the five simulations)

11.3 Summary

This brief analysis analysed three ways in which cities can improve the efficiency of their energy systems: improving the quality of the built environment, increasing the density of the urban layout, and using integrated, distributed energy systems, such as CHP. A few general conclusions can be drawn from this case study:

1. Cutting urban energy use by half is possible through integrated approaches that address the quality of the built environment (buildings efficiency), urban form and density, and urban energy-systems optimization (e.g. cogeneration).
2. Final energy use is not a sufficient indicator of energy system performance. In cogeneration systems in particular, this metric may show an increase in delivered fuel use that masks upstream conversion and distribution losses. This effect also occurs in bioenergy-based supply systems. Primary energy consumption should therefore be the basis of scenario comparisons.
3. Annual energy system costs (i.e. the costs of energy conversion and distribution), but not demand side measures, such as increased building efficiency or urban layout, are dominated by fuel costs.

However, these costs are distributed differently between stakeholders in each of the scenarios. In current practice, most of the capital and fuel costs will be paid by end consumers, whereas in a distributed energy system much more of the costs will be borne by energy utilities and project developers. This suggests that in order to achieve overall system efficiency, policy makers should design markets that help utilities to implement distributed energy installations despite their unique capital and fuel cost structures.

Increased urban density and improved building efficiency deliver the primary energy consumption and carbon emission savings of about one-third each; distributed energy systems provide an approximately 10 per cent primary energy and carbon emissions saving. This indicates the importance of urban planning measures. These decisions – for example, on the building energy performance standards or on the location of infrastructure – are difficult to change in retrofit and can lead to significant increases in energy consumption; in the cases studied here, urban sprawl led to a one-third increase in primary energy consumption. Efficient distributed energy systems can, to a certain extent, be retrofitted into existing urban forms but they too can benefit from long-sighted urban planning by encouraging sufficient demand density and by reducing the costs of network infrastructure.

The above simulations suggest that the effects on energy demand of urban form and density, and that of the energy efficiency characteristics of technologies, processes and practices (e.g. buildings) are of comparable magnitude; that is, they offer comparable size 'mitigation wedges' to paraphrase a concept developed by Pacala and Socolow (2004). Conversely, the impact of narrow energy systems optimization (e.g. through cogeneration of renewables) is much smaller.

12

Urban air quality management

Shobhakar Dhakal

Environmental constraints already form a significant dimension of national energy policies with issues such as acid rain, climate change, impacts of dams and others on top priority. For the urban system, it plays an even greater role given the predominance of fossil fuels as source of energy and resulting air pollution in cities where activities are concentrated due to the denser settlements.

12.1 Air pollution trends

Energy and air pollution are closely linked because air pollution is primarily a result of fuel burning in power generation, industries for domestic production and exports,[1] transport, and commercial and residential sectors. Low-quality fuels, such as coal, biomass, and high-sulfur diesel, emit more air pollutants than cleaner energy sources. From the urban energy-usage perspective, the literature shows that high urban density tends to be associated with lower per capita energy uses (see Chapter 9), which reduces air-pollution problems somewhat. However, the trade-off at this density reflects on the issue of air-pollution control. High density makes air-pollution control more urgent and requires better management systems, especially in the rapidly growing dense and large Asian megacities.

Historically, the concentrations of pollutants such as SO_2 and total suspended particles (TSPs), which mainly result from industrial production systems were concentrated in cities, have declined in industrialized cities. In the United States, the average national SO_2 level at 147 sites in 2007 was 0.0038 parts per million (ppm), which is 68 percent lower than 0.0118 ppm in 1980, while the National Ambient Air Quality Standard remains at 0.03 ppm (USEPA 2009). In Japan, SO_2 average levels have declined from 0.06 ppm in 1967 to 0.017 ppm in 1980, and further by 50 percent in 1980–2005 (MOE 2005). A key component of industrial air-pollution control mechanism was the development and deployment of end-of-pipe technologies (such as flue-gas desulfurization and particulate removal), and the introduction of cleaner fuels under stringent air-quality legislation. The 'pollute first and clean up later' paradigm had actually worked for many industrial cities, but one generation suffered acute air-pollution problems. Currently, the SO_x and TSP pollutants in industrialized countries are no longer an issue for energy consumption in cities, because urban energy systems are dominated by electricity or other cleaner fuels, and emissions from large point sources are tightly controlled.

In developing countries, SO_x and TSPs have shown a decreasing trend in recent years (Schwela et al. 2006; CAI-Asia, 2010) for selected Asian cities. SO_2 is reasonably within limits in many cities, while TSPs remain far higher. Yet, a recent survey of monitoring data of a greater sample of 213 Asian cities, for 2008 shows that 24 percent of cities' (mostly Chinese cities) annual average SO_2 concentrations do not even meet WHO's 2005 guideline for twenty-four-hour average[2] (Figure 12.1). Today, cities in developing countries can learn from the experiences of the industrialized countries and have access to the technologies developed previously to tackle SO_x and TSPs. In contrast to the developed countries, industrial relocation and FDIs constitute a key aspect of air-pollution control in the developing countries, with both positive and negative effects. At the same time, relocation of existing dirty industries of cities from populated areas to less populated areas or outside the cities has contributed to reductions in industrial air pollutants. Given that pollution-control technologies are readily available, the industrial air pollution in developing countries today largely results from either an inability to pay for the technology or inherent institutional, policy, and market problems.

While industrial air pollutants are falling, the challenges to control pollutants from mobile sources, as a result of automobile dependency of cities, in particular for particulate matter (PM), NO_x and O_3, are increasing. Even in the cities of industrialized nations, to reduce the levels of NO_x, O_3, and fine particles is proving a challenge. In the United States, monitored data show that the average levels of O_3 in 269 sites (0.078 ppm in 2007 – a reduction of 21 percent from 0.10 in 1980 – eight-hour average) slightly exceed National Air Quality Standards (USEPA 2009). A recent report by the American Lung Association (ALA 2009) shows that 125 million people (42 percent) in the United States live in counties that have unhealthy levels of either O_3 or particle pollution (PM2.5). In Japan, the compliance rate of monitoring sites for O_3 is extremely low – merely 0.2 percent in 2004 (MOE 2005).

The WHO estimates show that over 500,000 deaths in 2000 result from outdoor air pollution in Asia alone which is about two thirds of the global death burden attributed to air pollution (WHO, 2002; HEI, 2010a). HEI (2010a) shows that a 10–µg/m3 increase in PM10 concentration was associated with an increase of 0.6 percent in the daily rate of death from all natural causes in Bangkok, Hong Kong, Shanghai, and Wuhan.

PM10 is one of the key public health issues in the cities of many developing countries, where their levels are many times higher than the WHO or USEPA guidelines (Figures 12.1 and 12.2 and Table 12.1). Only 160 million people in cities worldwide are breathing clean air, more than 1 billion need improved urban air quality, and for 740 million urban dwellers air quality is above the minimum WHO limits. Figure 12.1 show that PM10 reduction is a struggling endeavor in Asian cities,

Average of Annual Average Ambient AQ in Selected Asian Cities (1993–2009)

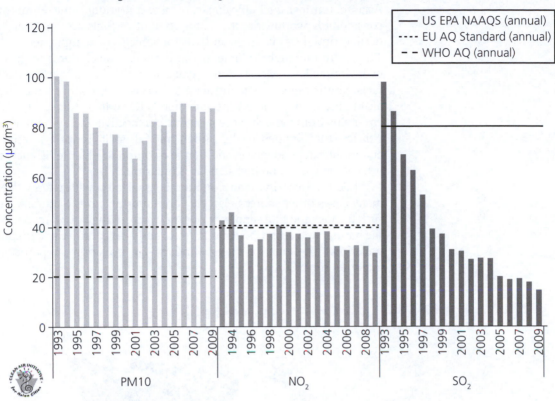

AQ = air quality; μg/m³ = micrograms per cubic meter; US EPA = United Stated Environmental Protection Agency; NAAQS = National Ambient Air Quality Standards; EU = European Union; WHO = World Health Organization; AQG = air quality guidelines; PM10 = Particles with particle diameters of 10 micrometers or less; NO_2 = Nitrogen dioxide; SO_2 = Sulfur dioxide.

Figure 12.1 Trends of major criteria air pollutants (1993–2009) for selected Asian cities
Source: CAI-Asia (2010); 243 Asian cities included in survey

despite considerable efforts. The ambient concentration levels remain, nonetheless, above WHO, USEPA, and EU limits in numerous cities worldwide (Figure 12.2 and Table 12.1).

A city-by-city analysis (Figure 12.3) further shows that the average of annual average PM10 in 230 selected Asian cities remains 4.5 times higher than the WHO Guidelines in 2008 (CAI-Asia 2010). Only two cities seem to be within the WHO Guidelines of 20μg/m³, while 58 percent of cities exceed the WHO Interim Target-1 (IT-1) of 70μg/m³. However, NO_2 levels are lower (Figure 12.3) and well within WHO Guidelines of 40 μg/m³ in many of the 234 cities that monitor NO_2 data and were surveyed (CAI-Asia 2010). However, NO_2 pollution yet seems to be of special concern for megacities (Schwela et al. 2006). The main reason for PM10 and NO_2 emissions is the rising number of private automobiles in cities of the developing world as a result of rising affordability, rising mobility demand, and slow development of public transport infrastructures. Over the years, the air quality standards have tightened (see CAI-ASIA 2010, for the latest Air Quality Standards for Asian countries), fuel efficiencies of new automobiles have improved considerably worldwide, the sulfur content of fuels are on a constant decline, and the emission standards for vehicles have tightened (Figure 12.4), but the high volume of automobile travel demand has far overwhelmed vehicle-efficiency gains and the impacts from cleaner fuels. Shifting engine size further affects this. This confirms that urban air-quality management especially needs to address transport-related emissions from a much more systemic perspective, including transport policies that influence the urban modal split toward a reduced automobile dependence in addition to traditional vehicle efficiency, and exhaust emission and fuel standards measures.

While the majority of air pollution is associated with energy use, in many cases other sources also play an important role. Natural factors, such as dust and fine sand particles, flow across the boundary between the natural and anthropogenic sources, and also contribute to urban air pollution. The role of transboundary air pollution is particularly important for SO_x.

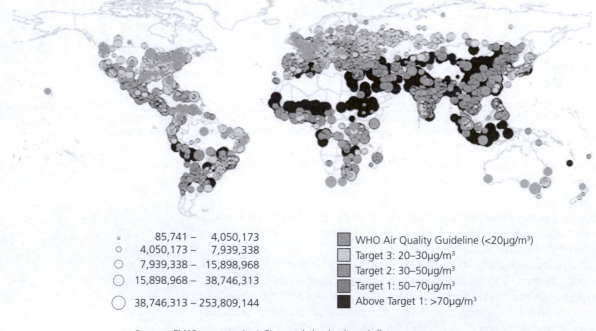

∘ 85,741 – 4,050,173	☐ WHO Air Quality Guideline (<20µg/m³)
○ 4,050,173 – 7,939,338	☐ Target 3: 20–30µg/m³
○ 7,939,338 – 15,898,968	☐ Target 2: 30–50µg/m³
○ 15,898,968 – 38,746,313	☐ Target 1: 50–70µg/m³
○ 38,746,313 – 253,809,144	■ Above Target 1: >70µg/m³

Exposure PM10 concentration* City population (capita.µg/m³)
Size of circle indicates exposure (Quintiles)
Color of circle indicates underlying PM10 Concentration (µg/m³) range: 7–358µg/m³

Figure 12.2 Human risk exposure to PM10 pollution in 3,200 cities worldwide. (For a numerical summary, see Table 12.1) (See color plate 15)
Sources: Doll (2009), Doll and Pachauri (2010), based on World Bank data

Table 12.1 Number of cities and residing population categorized by ambient PM10 WHO air quality standards (ACQ = WHO air quality guidelines met (less than 20 micrograms per cubic meter), for definition of Target 1 to 3 concentration standards, see Figure12.2); for a sample 3,200 cities globally and by three regions (ALM=Africa, Middle East, Latin America and the Caribbean)

Global	# of Cities	Population (millions)
ACQ	446	164
Target 3	809	385
Target 2	777	409
Target 1	362	260
Above Target 1	803	739
Annex-1	**# of Cities**	**Population (millions)**
ACQ	325	121
Target 3	610	314
Target 2	371	183
Target 1	51	41
Above Target 1	26	12
ALM	**# of Cities**	**Population (millions)**
ACQ	115	41
Target 3	160	60
Target 2	228	126
Target 1	132	103
Above Target 1	205	160
Asia	**# of Cities**	**Population (millions)**
ACQ	6	2
Target 3	39	11
Target 2	178	101
Target 1	179	116
Above Target 1	572	567

Sources: Doll (2009) Doll and Pachauri (2010), based on World Bank data

Figure 12.3 Urban concentrations of PM10 and NO$_2$ concentration in Asian cities (2008). (See color plate 16)
Source: CAI-Asia (2010)

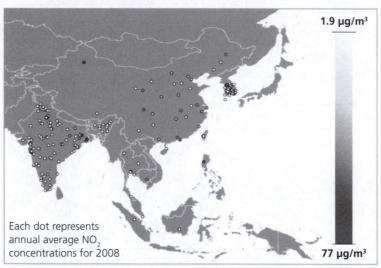

Each dot represents annual average NO$_2$ concentrations for 2008

Note: Annual NO$_2$ concentrations range from 1.9 (minimum) to 77 (maximum) micrograms per cubic meter (µg/m³)

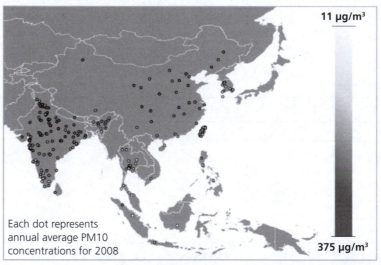

Each dot represents annual average PM10 concentrations for 2008

Note: Annual PM10 concentrations range from 11 (minimum) to 375 (maximum) micrograms per cubic meter (µg/m³). PM10 = Particles with particle diameters of 10 micrometers or less.

12.2 Examples of air pollution control measures and urban energy

A wide variety of air-pollution control policies and measures are in place globally. Some are system-wide and comprehensive measures, while others address a specific sector or technology, depending on the prevailing sources of the air pollutants. Some of these measures are regulatory, while others are technological, managerial, or a mixture. Here a few representative examples are illustrated that touch on a broad range of such measures, namely legislation, market, court rulings, and technology. Each has different implications for urban energy systems.

Country	95	96	97	98	99	00	01	02	03	04	05	06	07	08	09	10	11	12	13	14	15	16	17	18
European Union	E1	Euro 2					Euro 3				Euro 4			Euro 5				Euro 6						
Hong Kong, China	Euro 1		Euro 2				Euro 3					Euro 4			Euro 5									
South Korea												Euro 4			Euro 5									
China^a						Euro 1				Euro 2			Euro 3			Euro 4								
China^e					Euro 1				Euro 2			Euro 3		Euro 4			Euro 5							
Taipei, China					US Tier 1										US Tier 2 Bin 7^f									
Singapore^a	Euro 1						Euro 2																	
Singapore^b	Euro 1						Euro 2					Euro 4								Euro 5				
India^c										Euro 2				Euro 3										
India^d				E1	Euro 2					Euro 3				Euro 4										
Thailand	Euro 1						Euro 2			Euro 3						Euro 4								
Malaysia		Euro 1											Euro 2^g											
Philippines									Euro 1				Euro 2								Euro 4			
Vietnam													Euro 2									Euro 4		
Indonesia											Euro 2													
Bangladesh^a											Euro 2													
Bangladesh^b											Euro 1													
Pakistan															Euro 2^a			Euro 2^b						
Sri Lanka										Euro 1														
Nepal				Euro 1																				

Notes:
*The level of adoption vary by country but most are based on the Euro emission standards
a – gasoline; b – diesel; c – Entire country; d – Delhi, Mumbai, Kolkata, Chennai, Hyderabad, Bangalore, Lucknow, Kanpur, Agra, Surat, Ahmedabad, Pune and Sholapur; Other cities in India are in Euro 2; e – Beijing [Euro 1 (Jan 1999); Euro 2 (Aug 2002); Euro 3 (2005); Euro 4 (1 Mar 2008); Euro 5 (2012)], Shanghai [Euro 1 (2000); Euro 2 (Mar 2003); Euro 3 (2007); Euro 4 (2010)] and Guangzhou [Euro 1 (Jan 2000); Euro 2 (Jul 2004); Euro 3 (Sep–Oct 2006); Euro 4 (2010)]; f – Equivalent to Euro 4 emissions standards; g – for gasoline vehicles only

Figure 12.4 Overview of vehicle emissions standards, Europe versus Asia
Source: CAI (2009)

12.2.1 United Kingdom 'smoke control area' regulation

The United Kingdom started air-pollution control with a strictly source-control approach. It gradually shifted to a complex, but integrated, and risk management effects-based approach (Longhurst et al. 2009). Intense pollution from domestic coal use persisted in the United Kingdom until the 1950s and 1960s. Heightened concerns after London's Great Smog episode of December 1952 led to the introduction of the Clean Air Act of 1956 as an emergency measure. Significantly, some key industrial cities had already taken pre-emptive action, but the politics of a national measure proved more difficult. The Clean Air Act enabled local authorities to control pollution by declaring Smoke Control Areas ('smokeless zones', in which the burning of coal was banned) to whole or part of the district. Various measures were also used to ease compliance with the regulation, such as subsidies for furnace switching and generous off-peak electricity tariffs. In this regime, each local authority publicized the fuels that could be used and a list of exempt appliances. The Clean Air Act was further extended in 1968 to address the question of unevenness in the implementation, because in wealthier cities it was progressive while in other cities the implementation was less than that desired. The major feature of this regime was to induce a shift from coal to electricity, natural gas, and other cleaner forms of energy and implied a major transformation in energy end-use patterns and systems. However, with increasing levels of transport-related pollution in UK cities, NO_x concentration levels can be high and close to the statutory limits. This, in turn, may restrict further expansion of CHP systems in urban areas.

12.2.2 The regional clean air incentives market

The California South Coast Air Quality Management District (SCAQMD) has used a market-based system since January 1994, known as RECLAIM (Regional Clean Air Incentives Market), to reduce system-wide air pollution. SCAQMD covers 10,473 square miles. At the launching of RECLAIM, it was expected to reduce the cost of achieving the same emissions reduction through a traditional command-and-control approach by some 40 percent (Harrison 2004). Under this system, each facility participating in the RECLAIM program (facilities that emit 4 tons/year or more of NO_x and/or SO_x in 1990 or any later years) are allocated RECLAIM trading credits (RTCs) equal to their annual emission limits for SO_x and NO_x. The facilities must meet these allocated emissions limits. If a facility can reduce emissions more than required, they can sell surplus trading credits to the credit market, which can be bought by facilities that could not meet their own emission-reduction targets. The system is designed in such a way that the total allowable emission cap was reduced over time until 2003, after which it remained stable. It requires facilities to cut their emissions by certain amounts each year. By 2003, the program anticipated reducing the total emission load for NO_x and SO_x in SCAQMD by 70 percent and 60 percent, respectively (Anderson and Morgenstern 2009). In January 2005, RECLAIM decided to reduce the emissions further, by 7.7 tons/day or about 20 percent by 2011 (see www.aqmd.gov/RECLAIM/index.htm) (Figure 12.5).

12.2.3 Introduction of compressed natural gas vehicles in New Delhi

To address high concentrations of air pollutants in Delhi, the Supreme Court of India, acting on public interests litigation filed at the behest of civil society, directed the government of the National Capital Territory of Delhi and other authorities in Delhi to act on mitigating air pollution through specific technological interventions. The legal deliberations began as early as 1990, but the key court decisions were made in 1998. The cornerstone was the conversion of all diesel public transportation into compressed natural gas (CNG) vehicles, which began implementation in April 2001. Along with the implementation of these directives, authorities enacted a series of other measures, such as expansion of the scope of CNG coverage, scrapping of older vehicles, improving fuel quality, implementing stringent emission standards for vehicles, and improvements in the infrastructures. The implementation involved a series of policy instruments, such as penalties for non-complying vehicles, sales-tax exemption, interest subsidy on loans to replace three-wheelers, making CNG-retrofitting kits available, expansion of CNG fueling stations, and others. The literature shows an ambiguous picture of the impact of CNG conversion on air pollution,

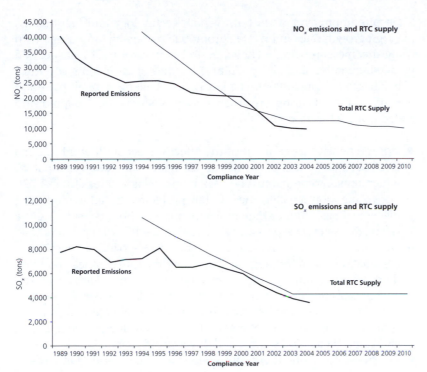

Figure 12.5 RECLAIM's impacts on emissions in SCAQMD area of California on NO$_x$ (left) and SO$_x$ (right) emissions versus allowable emission cap (RTCs) (in tons)

Source: Anderson and Morgenstern (2009)

but generally agrees that CNG conversion was one of the several key factors, and that it triggered other measures that led to improvements in the air pollution of Delhi (Goyal 2003; Kathuria 2004; Jalihal and Reddy 2006; Ravindra et al. 2006; Kandlikar 2007; UNEP 2009).

12.3 Key policy issues

The ways and methods adopted to mitigate urban air-pollution impacts in urban energy systems differ widely. In Delhi, deliberate CNG introduction to reduce air pollution created a new urban energy supply, demand, and infrastructure, largely following a technology strategy. The ban on coal-fired boilers in Chinese cities such as Beijing and Shanghai led to greater use of electricity and natural gas (Dhakal 2004). As an example of a regulatory approach to urban air pollution, it largely follows the historical United Kingdom 'smokeless zone' regulatory model. City energy-system decisions in China are influenced by air-pollution mitigation, public transport improvements, and energy-security concerns (Dhakal 2009). The will to control PM10 prompted many cities in Asia, America, and Latin America to move progressively toward discouraging diesel. Europe is moving more on the path of dieselization with stricter control of the sulfur content of diesel combined with particulate filters in automobiles.

Despite improvements in both vehicle technology and fuel quality, the high growth rates in private automobile ownership and usage with rising income is proving a challenge to the control PM10, suspended particulate matter, and NO$_x$ pollutants in cities of developing countries (Dhakal and Schipper 2005). The key policy challenges for air pollution that have direct bearing on urban energy systems are (adapted and added from Schwela et al. 2006):

- Comprehensive assessment of the effectiveness of different options is needed in cities, but requires adequate institutional capacity, which remains comparatively weak in many large cities (Table 12.2), not to mention smaller ones. Often problems are addressed on a piecemeal basis and at the end of the pipe without considering the complete systemic aspect and thus rebound effects are prevalent.
- Development of more reliable inventories of air-pollution emissions is essential. Cities are not regularly updating their inventories and often there are serious ambiguities in emission volume and sources data. The energy data and information for drivers that are linked to energy and air pollution are on short supply. The capacity to update and improve continuously on such inventories and the ability to analyze the changing landscape of emission sources and drivers needs to build.
- The need to adopt more stringent vehicle-emission standards is evident. The pace of adopting new emission standards in the face of rapidly rising private transportation is very slow. In addition, a reasonable global harmonization of air-quality and technology standards is needed. Currently, decision makers are torn between the Euro Standards and the USEPA standards, which affect technology choice and fuel regimes differently.
- Introducing cleaner fuel more actively for motor vehicles, industries, and power plants is necessary.

Table 12.2 Classification of urban air-quality management capacity in Asian cities

Capability Classification	Cities
Excellent I	Hong Kong, Singapore, Tapei, Tokyo
Excellent II	Bangkok, Seoul, Shanghai
Good I	Beijing, Busan
Good II	New Delhi
Moderate I	Ho Chi Minh City, Jakarta, Kolkata, Metro Manila, Mumbai
Moderate II	Colombo
Limited I	Hanoi, Surabaya
Limited II	Dhaka, Kathmandu
Minimal	-

Source: Schwela et al. (2006)

- Transport polices that affect urban mobility choices need to complement vehicle- and fuel-specific policy measures. In their absence, any air-quality improvements are likely to be quickly overwhelmed by continued motorized transport growth.
- Despite good policies on technology and fuel, inadequate emphasis on inspection and maintenance of systems remains one of the key challenges. Much of the existing air-pollution problems can be addressed by simple implementation and stricter enforcement of existing legislations, standards, and inspection and monitoring regimes for air quality.
- For transboundary air-pollution issues, such as acid rain and black carbon (emerging as key problems, particularly in Asia), regional approaches and regimes are needed, but such regional coordination is emerging only very slowly in many world regions.
- To harmonize many environmental issues within common policy responses, estimation of the co-benefits of air-pollution management with respect to human health, urban energy system improvement, energy security, climate change mitigation, and ecosystems in general is essential. In developing countries such a co-benefit approach can help devise limited resources more efficiently, and also broaden the technology and financial resources available for air-pollution control.

Notes

1   Global market integration and economic structural change can result in relocations of industrial activity, exercising an additional effect on local pollution.
2   WHO carried out an update of its Air Quality Guidelines in 2005. For SO_2, this is 20 $\mu g/m^3$ for 24-hr-average. Guideline for annual average concentration is not established. but it would be much lower than 24-hr-average.

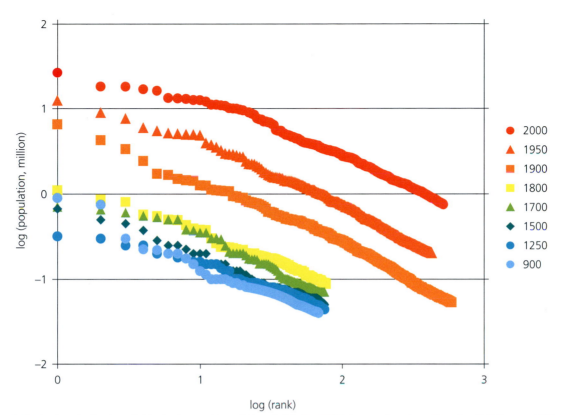

Plate 1 Rank–size distribution of cities in the world, AD 900 to 2000. Pre-1950 data are based on Chandler (1987), post-1950 on UN DESA (2010). Sample size varies between fewer than 80 in the pre-1800 period to close to 600 cities in the post-1900 period

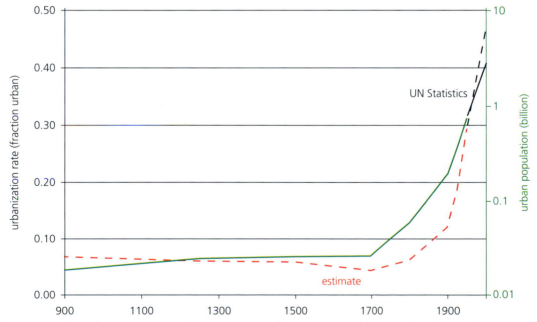

Plate 2 Estimates of global urban population (billion, right axis, solid green lines) and fraction urban (fraction, left axis, dashed red lines) AD 900–1950 inferred from Chandler (1987), and UN DESA (2010) data 1950 to 2000 (black lines)

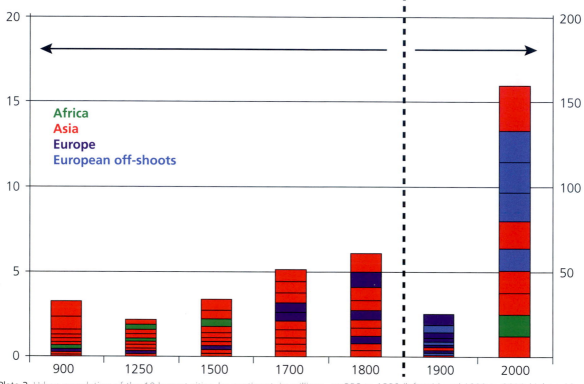

Plate 3 Urban population of the 10 largest cities, by continent, in millions, AD 900 to 1800 (left axis) and 1900 to 2000 (right axis)
Sources: data: Chandler (1987); UN DESA (2010)

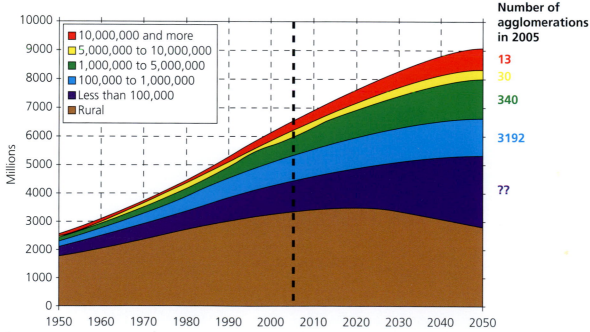

Plate 4 Population by residence and settlement type (millions). Historical (1950–2005) and projection data (to 2050) (for 2005 statistics, see Table 2.3)
Source: adapted from UN DESA (2010)

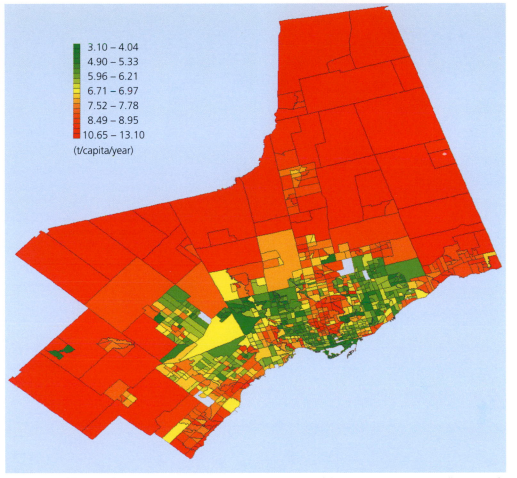

3.10 – 4.04
4.90 – 5.33
5.96 – 6.21
6.71 – 6.97
7.52 – 7.78
8.49 – 8.95
10.65 – 13.10
(t/capita/year)

Plate 5 Total GHG emissions from Toronto (tons CO_2–equivalent/capita/year). High-resolution images as well as maps for various energy-demand subcategories (residential, transport, etc.) are available from VandeWeghe and Kennedy (2007)

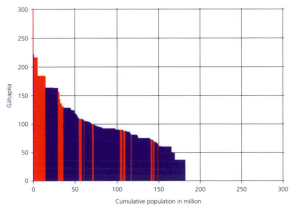

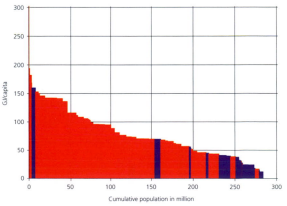

Plate 6 Per capita (direct) final energy consumption (TFC) (GJ) versus cumulative population (millions) in urban areas (*n*=160) of Annex-I (industrialized) countries. Red indicates urban areas with per capita TFC *above* the national average. Blue indicates per capita TFC *below* the national average

Plate 7 Per capita (direct) final energy consumption (TFC) (GJ) versus cumulative population (millions) in urban areas (*n*=65) of non-Annex-I (developing) countries. Red indicates urban areas with per capita TFC *above* the national average. Blue indicates per capita TFC *below* the national average

Plate 8 Example of assessing local renewable potentials: suitable roof-area identification for solar PV applications for the city of Osnabrück, Germany. Red: roof area well suited for PV; orange: suitable; yellow, only conditional suitability for PV applications; gray: shadowed roof area (unsuitable)
Source: modified from Ludwig et al. (2008)

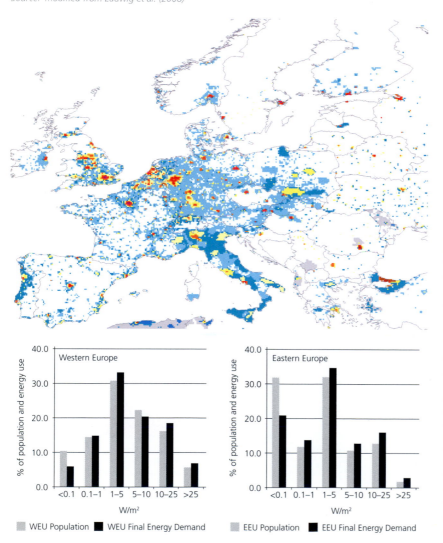

Plate 9 (**Top**): Spatially explicit energy demand densities in Europe (W/m²): blue and white areas indicate where local renewables can satisfy local low-density energy demand (<0.5–1 W/m²); yellow, red, and brown colors denote energy demand densities above 1, 5, 10, and 25 W/m² respectively.

(**Bottom**): Distribution of population (gray) and final energy demand (black) (in percent) as a function of energy demand density classes in W/m² for Western Europe (left panel) and Eastern Europe (right panel). Only 21% (Western Europe) and 34% (Eastern Europe) of energy demand is below an energy demand density of 1 W/m² amenable to full provision by locally available renewable energy flows. The high energy densities of cities require vast energy 'hinterlands' that can be 100–200 times larger than the territorial footprint of cities proper requiring long-distance transport of renewable energies

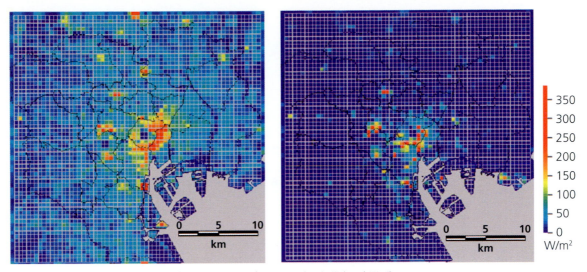

Plate 10 Sensible (left) and latent (right) anthropogenic heat emission in Tokyo (W/m²)
Source: Ichinose (2008)

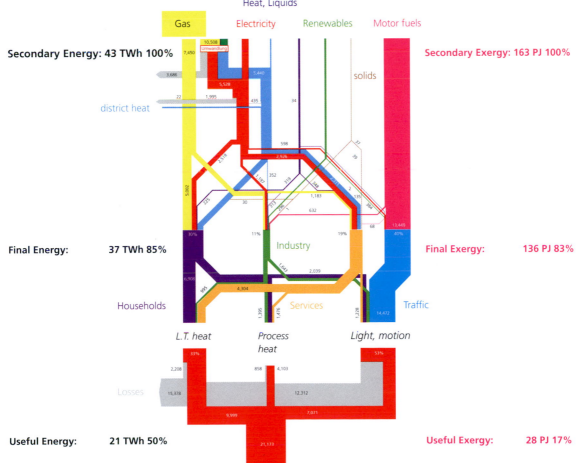

Plate 11 Energy and exergy flows in the City of Vienna in 2007 between secondary and useful energy/exergy
Sources: Energie Wien (2009) (approximate) exergy efficiencies based on Gilli et al. (1996)

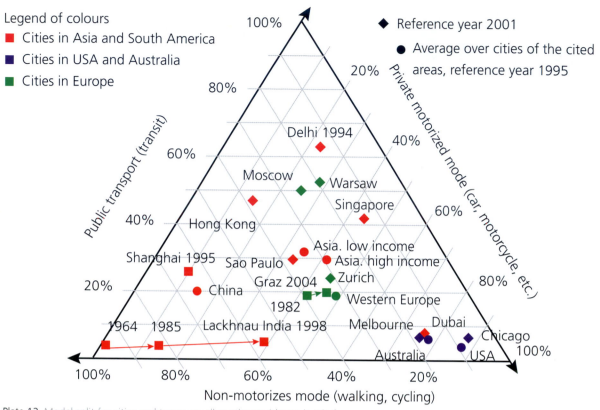

Plate 12 Modal split for cities and towns on all continents (shares in trips)
Sources: Kenworthy and Laube (2001); Padam and Singh (2001); Zhou and Sperling (2001)

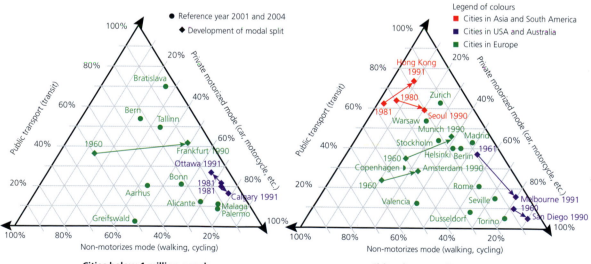

Cities below 1 million people

Cities above 1 million people

Plate 13 Modal split of journeys to work in medium-sized towns with a population below 1 million people (left panel) and in cities with a population above 1 million people (right panel) in high-income economies for reference years 2001 and 2004 and selected time trends since 1960
Sources: Urban Audit (2009); Vivier (2006); Steingrube and Boerdlein (2009); Wapedia (2009)

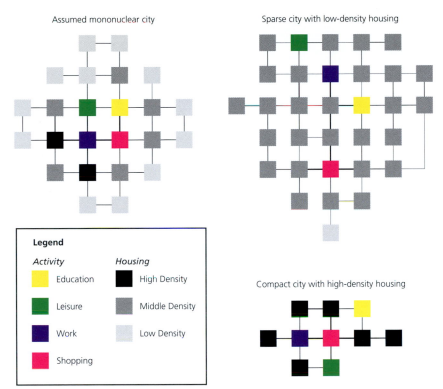

Assumed mononuclear city

Sparse city with low-density housing

Legend

Activity

🟨	Education
🟩	Leisure
🟪	Work
🟥	Shopping

Housing

⬛	High Density
⬜	Middle Density
⬜	Low Density

Compact city with high-density housing

Plate 14 Urban layouts (left to right): the assumed mononuclear city, a compact city with high-density housing, a sparse city with low-density housing. In each figure, the coloured cells represent activity provision: green for leisure, blue for work, pink for shopping, and yellow for education. The grey cells represent housing at different densities, with the labels indicating the density in dwellings per hectare. The black lines connecting the cells indicate road connections and indicative traffic flows

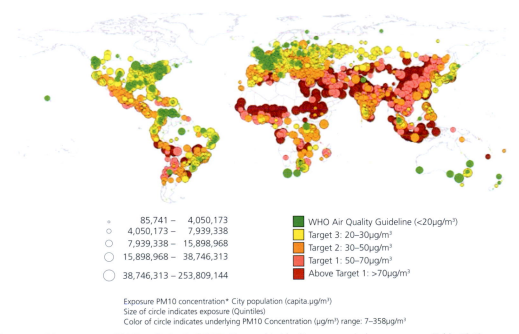

∘	85,741 – 4,050,173
○	4,050,173 – 7,939,338
○	7,939,338 – 15,898,968
◯	15,898,968 – 38,746,313
◯	38,746,313 – 253,809,144

🟩	WHO Air Quality Guideline (<20µg/m³)
🟨	Target 3: 20–30µg/m³
🟧	Target 2: 30–50µg/m³
🟥	Target 1: 50–70µg/m³
🟥	Above Target 1: >70µg/m³

Exposure PM10 concentration* City population (capita.µg/m³)
Size of circle indicates exposure (Quintiles)
Color of circle indicates underlying PM10 Concentration (µg/m³) range: 7–358µg/m³

Plate 15 Human risk exposure to PM10 pollution in 3,200 cities worldwide. (For a numerical summary, see Table 12.1)
Sources. Doll (2009), Doll and Pachauri (2010), based on World Bank data

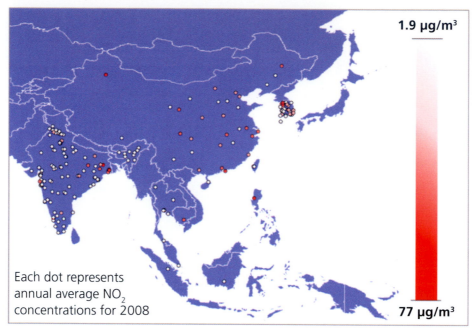

1.9 µg/m³

77 µg/m³

Each dot represents annual average NO₂ concentrations for 2008

Note: Annual NO₂ concentrations range from 1.9 (minimum) to 77 (maximum) micrograms per cubic meter (µg/m³)

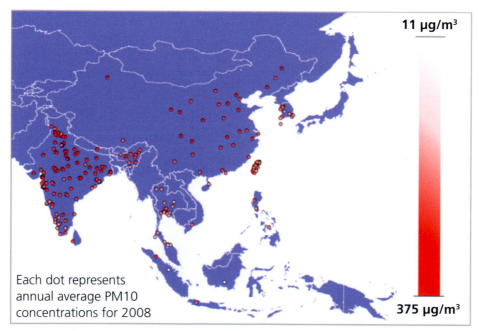

11 µg/m³

375 µg/m³

Each dot represents annual average PM10 concentrations for 2008

Note: Annual PM10 concentrations range from 11 (minimum) to 375 (maximum) micrograms per cubic meter (µg/m³). PM10 = Particles with particle diameters of 10 micrometers or less.

Plate 16 Urban concentrations of PM10 and NO₂ concentration in Asian cities (2008).
Source: CAI-Asia (2010)

13

Summary and conclusion

Arnulf Grubler and **David Fisk**

13.1 An urbanizing world

This book has worked with 'urban' as a designation of those settlements where basic services such as water and drainage and food distribution are provided by collective services. Defined in this basic way the world is already predominantly urban. More than 50 percent of the global population lives in such settlements. In industrially developed countries the fraction that is urbanized has risen to 80–90 percent. Because industry and commerce are often the *raison d'être* of clusters of population, it is no surprise that urban areas account for the larger share of a nation's economic activity. (This book estimates that some 80 percent of the world's GDP is generated in urban areas.) Urban areas will also absorb almost all the global population growth to 2050, amounting to some three billion additional people. Urban populations were once self-limiting because of the high risk of epidemics among so many crowded in so small a space. But with access to clean water and sanitation and modern health care, this limit is no longer significant. Also, as with the European Industrial Revolution, there is significant migration from rural to urban areas. Over the next decades the increase in rural population in many developing countries will be overshadowed by population flows to the cities. Rural populations globally, are expected to peak at a level of 3.5 billion people by around 2020 and decline thereafter. This global view of course obscures important differences in projected regional trends. Whereas rural populations in Asia are projected to decline rapidly after 2020, the African rural population is projected to continue to grow at least to 2040, before also declining. Shrinking cities in the developed world are an increasing phenomenon in urban dynamics, and could continue as below-replacement fertility levels outstrip increased longevity and so lead to declining populations in almost all high-income countries (and potentially in presently low-income countries in the long term as well). The impacts of population contraction on urbanization remain a major unknown. However, whatever the detail, it is still clear that the future is urban. That then is where the policy formulation arena should be.

Patterns of urban settlement growth have been and will remain heterogeneous. Most of the growth will continue to occur in small- to medium-sized urban centers. Despite much public attention, the contribution of 'megacities' to global urban-population growth will remain comparatively small. This pattern is consistent with the remarkable robustness of the distribution of city-size classes over time and across different regions. Each settlement presents its own opportunities and challenges, but the expected vast growth in small- to medium-sized cities in the developing world is a particular challenge. Whereas urban centers provide unique opportunity to marshal resources for collective action, data and information to guide such action are largely absent in many smaller-scale cities, local resources to tackle development challenges are limited, and governance and institutional capacities can be weak. So an increasing urbanized world may require a substantial reconfiguration of existing governance structures to address twenty-first century challenges. Not least among these is the provision of energy that is affordable, secure and clean.

13.2 Urban energy use

Energy use is prerequisite of the urban settlement. Without it goods and services could not be transported to the citizen, nor work done, nor living spaces conditioned. Ensuring the provision of 'affordable, secure and clean' energy (i.e. energy *supply*) has been traditionally the role of central government. But how energy is *used* (i.e. energy *demand*) is likely to be influenced greatly by more local decisions. For example the density of energy use also often determines local environmental quality and as a consequence local government takes an interest in the location and operation of energy consuming processes. But in many cases the balance between local and central government may not be best suited for the challenges ahead and the investment required. Patterns of energy demand may become as important in the future for successful settlements as patterns of energy supply. This will be true whether the policy maker is developing policies to improve national energy security, national competitiveness or as part of an international effort to reduce greenhouse gas emissions. All these policies will have to entail an important energy demand component that in many instances is framed by local decision making processes at the urban scale.

Accounting for energy use as a precursor to evidence-based policy formulation, can be notoriously sensitive to the choice of systems boundaries. Urban energy policy is no exception. In some respects it is even more sensitive because few modern cities are self-contained economic units. Thus estimating the urban share of current world-energy use varies as a function of assumed system boundaries in terms of spatial scales (cities versus agglomerations), energy-systems definition (final commercial, total final, and total primary energy), and the boundary drawn to account for embodied energy in a city's goods and services, both imported and exported. Since there is so much at stake there is always a risk that political rhetoric may exploit fungible boundaries. Sound and transparent analytic techniques are therefore essential for a quality political debate.

The direct use of national energy-reporting formats to the urban scale is often referred to as a 'production' approach. This contrasts to a 'consumption' accounting approach that pro-rates associated (primary) energy uses per unit of expenditure of urban consumer expenditures, thus accounting for energy uses irrespective of their form (direct or embodied energy) or location (within or outside a city's administrative boundary). Both approaches provide valuable information. 'Production' and 'consumption' measures are complementary tools to inform urban policy decisions. However, because of their complexity, future urban studies need to adhere to high standards in terms of clarity and documentation of the terminology, methodology, and underlying data used. Specifically, to improve comparability, all accounts based on a consumption approach (which are data and time intensive to prepare, and so exist only for a very limited set of megacities) need to be complemented by corresponding production-based energy accounts (which are much simpler and easier to assemble by government authorities). This problem of viewpoint is highlighted in the recent importance attached to providing GHG emission inventories for cities. Because of the possible confusion as to their basis no urban GHG-emission inventory should be published without the underlying energy data used in the assessment.

Available estimates of current urban energy use based on a production approach (direct final energy use, or primary energy use, i.e. including pro-rated upstream energy sector conversion losses) suggest that urban energy use accounts for between 60 percent and 80 percent of global energy use. Mirroring the growing importance of urban areas in demographic and economic development, urban energy use will continue to grow further as a fraction of total global energy use. There is clearly a limit to the extent to which energy sustainability issues can be handled simply by action 'upstream' of the final energy delivered to the urban settlement. This limitation implies that future energy sustainability challenges will need to be addressed and solved primarily by action in urban settings. This is the challenge of energizing cities.

There is great heterogeneity in urban energy-use patterns, especially when manufacturing and transport energy uses are included. In many developing countries, urban dwellers use substantially *more* final energy per capita than their rural compatriots. This primarily reflects ownership of energy consuming appliances and vehicles associated with higher urban incomes compared to rural incomes. Conversely, in many industrialized countries per capita final energy use of city dwellers including transport (i.e., based on a production-accounting approach) is often *lower* than the national average, which reflects the effects of compact urban form, settlement types (multi- versus single-family dwellings) and availability and/or practicability of public transport infrastructure systems compared with those in the suburban or rural sprawl. The few available data, however, suggest that urban energy use in high-income countries is not substantially different from the national average when using a consumption-based accounting approach that also includes energy embodied in imports into the urban settlement. So, the effects of lowered direct final energy use through a more service-oriented urban economy, urban form and density, and lower transport energy use, are largely offset by higher embodied energy use associated with the energy consumption of capital goods bought by the higher urban incomes. For countries with a large proportion of its citizens on low-incomes, the available data are too sparse to allow a similar comparison. However, it can be conjectured with some confidence that, because of the much higher income differential between urban and rural populations in low-income countries, a country's per capita urban energy use will be significantly higher compared to national averages in a consumption-based accounting framework as well. These arguments point to a powerful leading indicator of future developments to come with rising urbanization and income growth in the developing world. Urbanization driven by an economic imperative implies both an increase in consumption and a change in the pattern of consumption.

Drivers of urban energy use include geography and climate, resource availability, socioeconomic characteristics, degree of integration into the national and global economy (imports/exports), and urban form and density. Not all of these can be influenced by local governance and decision making. Priorities for urban energy and sustainability policies, therefore, should focus where local decision making and funding also provides the largest leverage effects: urban form and density (which are important macro-determinants of urban structures, activity patterns, and hence energy use), the quality of the built environment (energy-efficient buildings in particular), urban transport policy (particularly the promotion of energy efficient and 'eco'-friendly public transport and non-motorized mobility options), and improvements in urban energy systems through cogeneration or waste-heat-recycling schemes, where feasible. Local action, however, also requires local capacities and responsibilities in addressing urban energy and environmental problems, including a mediating role among the multiple stakeholders characteristic of decentralized urban decision making.

Conversely, the promotion of local solar or wind renewables will, at best, have a marginal impact on the overall energy use of larger cities (typically <1 percent)[1] because of the significant energy–density mismatch between (high) urban energy use and (low) renewable energy flows per unit land area available in urban areas. Smaller cities, however, could provide more avenues to integrate renewable energy into urban energy systems than possible in large cities. Cities could also play an important role as consumers of renewable energies, creating niche market impulses as well as potentially exerting leverage on the application of sustainable social and ecological production criteria for their renewable energy suppliers.

Nonetheless, urban energy and climate policy should recognize that the most productive local decisions and policies influence the *efficiency* of urban energy use that is the demand side of the energy system, rather than its supply side.

13.3 Facing the twenty-first century challenges

The grand objective of public policies is often grandly stated as to increase social welfare. For energy policy this is frequently translated as energy that is 'affordable, secure and clean'. These three themes take on a special character when we realize that in the next decades they are largely to be delivered in an urban context.

13.3.1 Equitable cities: urban poverty reduction

Migration rates and urban population growth can overwhelm the provision of basic urban services to the poorest urban dwellers. Several hundred million urban dwellers in low- and middle-income nations lack access to electricity and are unable to afford cleaner, safer household fuels or lack access to affordable and safe public transport options. Most are in low-income nations in Southeast Asia and sub-Saharan Africa. In many low-income nations, more than half the urban population still rely on charcoal, fuelwood, straw, dung, or wastes for cooking, with significant adverse consequences for human health and local air quality. A large part of the poor urban population that lacks clean fuels and electricity earns so little that it cannot afford access, but even where it could, can face political or institutional obstacles for energy access. Innovations have reduced the costs of energy access for the poor – for instance, rising tariffs with low prices for 'lifeline' consumption, pay-as-you-use meters, and standard 'boards' that remove the need for individualized household wiring, but important political/institutional barriers remain.

The constraints on supporting the shift to clean fuels and providing all urban households with electricity are less to do with energy policy than with local authorities' handling of issues of informal settlements. A large part of the population that lack clean energy and electricity live in informal settlements. It is mostly in nations where this antagonistic relationship between local government and the inhabitants of such settlements has changed, through widespread public support to upgrade 'slums' and squatters, that clean energy and electricity reaches urban poor groups.

Housing, infrastructure, energy, and transport services are the key sustainability challenges to accommodate some three billion additional urban dwellers in the decades to come, especially in low-income countries. Informal settlements will often be one of the transitional forms of settlement for many of these new urban dwellers and will require a much more proactive, anticipatory policy approach, especially with respect to the location of informal settlements and subsequent infrastructure connections and upgrading programs.

Inevitably energy-wise, low-cost and fast implementation options will take precedence over 'grand' new urban designs that require unrealistically large sustained capital provision over long periods. In low-income countries access to clean cooking fuels and electricity, as well as pro-poor transport policies, which include safer use of roads by non-motorized modes (walking and bicycling) and making public transport choices available (e.g., through bus rapid transit (BRT) systems) need to receive more attention.

13.3.2 Livable cities: improving urban density and form

Urban density and form are not only important determinants for the functionality and quality of life in cities, but also for their energy use. Historically, the diversity of activities and ensuing economic and social opportunities that are the major forces of attraction to urban settings were provided by high density and co-location (mixed land-uses) of a diversity of activities that maximize the 'activity zone' of urban dwellers while minimizing transport needs. This urban history contrasts with recent decades of trends in developed countries toward lower urban densities, which include widespread urban sprawl, and even 'ex-urban' developments.

It is widely recognized that there is no theoretical or practical argument for defining a universal 'optimal' form or density for a city. In theory (and often also in practice) higher densities increase the *economies of scope* (i.e., of activity variety) in a city. These are the main locational attractions of urban places in terms of potential number of jobs, breadth and variety of specialized trades and economic activities, along with cultural and many other attractions, usually summarized in urban economics as *positive agglomeration externalities* that also extend to urban infrastructures (e.g., communication and transport networks). Conversely, higher densities can entail negative externalities as well, such as congestion, high land prices and rents that limit the quality of residential living space for urban dwellers, or environmental problems (noise, air pollution).

Nonetheless, empirical data strongly suggest that the *net balance* of these positive and negative agglomeration externalities remained stable for extended periods of time, as illustrated by the remarkable stability of the rank–size distribution of cities (with dominating positive or negative net agglomeration externalities the growth of the *ensemble*[2] of larger cities should be above or below that of smaller cities, which is not the case). The historical evolutionary processes that govern urban growth have played out differently over time and space, which results in *path dependency*. Cities that evolved along alternative pathways have alternative density levels from high-density 'Asian' (e.g., Tokyo, Shanghai, Mumbai) and 'old European' (e.g., London, Madrid, Warsaw) to low density 'new frontier' (e.g., Los Angeles, Brasilia, Melbourne) pathways, each of which have different structural options available to improve energy efficiency and optimize urban energy and transport systems in terms of sustainability criteria.

Despite this diversity, two important generalizations can be drawn. First are the implications of urban density on the requirements of urban energy systems that need to be basically *pollution free*, as otherwise even relatively clean energy forms can quickly overwhelm the assimilative capacity of urban environments. This especially applies to the million, decentralized energy end-use combustion devices (stoves, heating systems, vehicles) for which end-of-pipe pollution control is often not an option. Thus, in the long-term all end-use energy fuels burnt in urban areas need to be of zero-emission quality. This requires energy vectors from remote 'clean' plants, as exemplified by electricity or (possibly) hydrogen. Natural gas plays the role of the transitional fuel of choice in many urban areas. This 'zero-emission' requirement for urban energy transcends the customary sustainability divide between fossil and renewable energies, as even 'carbon-neutral' biofuels when used by millions of automobiles in an urban environment will produce unacceptable levels of NO_x or O_3 pollution.

Second, the literature has identified repeatedly important size and density thresholds for various energy conversion technologies that are useful guides for urban planning. The importance of these urban size/density thresholds extends to specialized urban infrastructures, such as underground (metro) transport networks that are, as a rule, economically (in terms of potential customers and users) not feasible below a threshold population size of less than one million. It also extends to energy (e.g., cogeneration-based district heating and cooling) and public transport networks, whose feasibility (both for highly centralized, and decentralized, distributed 'meso'-grids) are framed by a gross[3] density threshold between 50 and 150 inhabitants ha (5,000–15,000 people/km^2). Such density levels of 50–150 inhabitants/ha certainly do *not* imply the need for high-rise buildings, as they can be achieved by compact building structures and designs, both traditional and new, including town or terraced houses, while still allowing for open public (parks) or private (courtyard) spaces – but they do preclude unlimited (aboveground parking) spaces for private vehicles. Zoning and parking regulation, combined with public transport policies and policies that promote non-motorized transport modes and walkability thus constitute the essential 'building blocks' of urban energy efficiency and sustainability 'policy packages'.

13.3.3 Mobile cities: more functional and green urban transport

Urban transport is a key policy concern for its crucial importance to the very functionality of cities. Two fundamental observations need to guide urban transport policies:

First, on a sustainability metric there is a clear contradiction between growing private motorized transport and growing energy use and pollution. This contradiction relates not primarily to the technological artifacts *per se* (automobiles or scooters), but rather to the *organizational form* of their *usage* as privately owned vehicles with correspondingly low occupancy rates (and thus high energy/emissions per unit service delivered). Well-designed taxi systems or automobile-sharing schemes are excellent demonstrations that the (selective) use of automobiles as modes of individual transport when needed (as opposed to their preordained use despite congestion simply because of a lack of alternatives) can be reconciled with the prerogatives of an energy-efficient city. Conversely, along the same sustainability metric, non-motorized mobility (walking and bicycling) and public transport schemes that function well and have high occupancy rates are the options of choice for urban mobility and should receive corresponding priority.

Second, the inter-dependencies between demand and supply of transport infrastructure often create a 'vicious circle' for urban transport planning: more automobiles lead to congestion, which improved urban road infrastructures aim to alleviate. But more (road infrastructure) supply *induces* yet more demand (individual mobility and land-use changes) in an ever-spiraling 'rat race' of 'supply following demand' growth. An improved public policy paradigm for urban transport drawing on improved modeling techniques needs to break this cycle of rebound effects through integrated urban energy transport policies that deal with both demand and supply.

The wide variation in urban transport choices observed in the modal split in different cities illustrates that urban mobility patterns are not *ex ante* given, but rather result from specific choices of individuals and decision makers. Urban transport choices can be modified, if both a strong determination for sustainable transport policy *and* a corresponding wide public acceptance of the overall goals of such a policy exist. This wider acceptance of the overarching goals is also required to implement some individual measures (e.g., traffic calming, pricing schemes (for roads and parking, etc.)). Restrictive measures that limit individual mobility by automobiles need to be complemented by proactive policies that enhance the attractiveness of non-motorized and public transport choices, and 'soft' policy measures (e.g., fees, tariffs) also need to be complemented by 'hard' (i.e., infrastructural investment) measures. The overriding goal is to turn the often automobile-dependent 'vicious' policy cycle into favoring a 'virtuous' cycle of non-motorized and public transport choices.

One key measure in this context will be the progressive internalization of external costs of motorized private transport along with the provision of high-quality alternative public and non-motorized transport. Estimates for Europe suggest that these external costs are at least in the order of 6–10 (Euro) cents per passenger-km, which when fully internalized could double private motorized transport costs. Comparable estimates for low-income countries are not available, but given their generally much higher road-accident rates, external costs are likely to be even larger. However, accompanying measures and strong leadership will be needed to increase the political acceptability of this invariably unpopular measure. Recent experiences with the introduction of road prices and congestion charges in cities across a wide political spectrum suggest that such policy approaches are both feasible and have the ability to alter urban transport behavior.

Successful public transport systems require a dense public transport network and a high service frequency with short intervals, which are only feasible with a minimum threshold of urban density. A rule-of-thumb goal might be to have only urban settlements within easy walking (<500 m) access to a viable public transport service. Investments in public transport systems need to find an appropriate balance between improvements that are less capital intensive and faster to implement, and radical solutions. BRT systems are, therefore, a much more attractive option for many cities in low-income countries than are capital-intensive subway systems, even though, in the long term, the latter offer the possibility of higher passenger fluxes and greater energy efficiency. Often there is no contradiction between incremental versus radical public transport policy options: for instance, BRT can also be considered as a transitional infrastructure strategy to secure public transport 'rights of way', which offer subsequent possibilities for infrastructure upgrades, for example in putting light rail systems in BRT lanes. Many of the new urban settlements being formed easily meet these density targets. The policy issue is to exploit the advantages before 'lock-in' into a private transportation-led 'vicious' development cycle takes hold.

13.3.4 Efficient cities: doing more with less energy

From all the major determinants of urban energy use – climate, position in the global economy, consumption patterns, quality of built environment, urban form and density (including transport systems), and urban energy systems and their integration – only the final three are amenable to a policymaking context at the urban scale.

Systemic characteristics of urban energy use are generally more important determinants of the efficiency of urban energy use than those of individual consumers or of technological artifacts. For instance, the share of high occupancy public and/or non-motorized transport modes in urban mobility is a more important determinant of urban transport energy use than the efficiency of the urban vehicle fleet (be it buses or hybrid automobiles). Denser, multifamily dwellings in compact settlement forms with a corresponding higher share of non-automobile mobility (even without thermal retrofit) can use less *total energy* than low-density, single-family 'Passivhaus'-standard (or even 'active', net energy-generating) homes in dispersed suburbs deploying two hybrid automobiles for work commutes and daily family chores. Evidently, urban policies need to address both systemic and individual characteristics in urban energy use, but their different long-term leverage effects should structure policy attention and perseverance.

In terms of urban energy-demand management, the quality of the built environment (buildings efficiency) and urban form and density that, to a large degree, structure urban transport energy use are roughly of equal importance. Also, energy-systems integration (cogeneration, heat cascading) can give substantial efficiency gains, but ranks second after buildings efficiency and urban form and density, and associated transport efficiency measures, as shown both by empirical cross-city comparisons and modeling studies reviewed in this book.

The potential for 'static' energy-efficiency improvements in urban areas remains enormous, as indicated by corresponding urban exergy analyses that suggest urban energy-use efficiency is generally less than 20 percent of the thermodynamic efficiency frontier; this suggests an improvement potential of more than a factor five. Implementing efficiency improvements (including systemic measures) should therefore receive highest priority. In the built environment, there is considerable inertia despite the relatively low or even negative cost of GHG abatement through refurbishment of buildings. There is therefore a strong argument for prescription and regulation regarding building standards.

Conversely, the potential of supply-side measures within the immediate spatial and functional confines of urban systems is very limited, especially for renewable energies. *Locally harvested* renewables can, at best, provide 1 percent of the energy needs of a megacity and a few percentage points in smaller, low-density cities because of the mismatch between (high) urban energy demand density and (low) renewable energy supply densities. Without ambitious efficiency gains, the corresponding 'energy footprint' of cities that import large-scale, centralized renewable energies (biofuels or electricity) will be vast and at risk of producing 'collateral' damages caused by large-scale land conversions (e.g., soil carbon perturbations and albedo changes), and competition over food and water. Given that all city systems are subject to uncertainty and change, there is also a need for option-based design techniques that allow for city growth and technological evolution, and that avoid strongly path-dependent solutions and 'lock-in' into urban energy-supply systems based on current- or near-term renewable options that, ultimately, will be superseded by third and fourth generation renewable supply technology systems. Improved urban energy-efficiency leverages supply-side flexibility and resilience, and thus adds further powerful arguments for strategies that focus on urban energy-efficiency improvements.

Finally, *energy security* is a re-emerging issue for many cities because energy security declined over recent decades and, in an uncertain world, security concerns need to be integrated increasingly into urban energy policies and sustainability transition analysis. Better efficiency and improved energy systems integration will also benefit urban energy security, although a further assessment of energy security is beyond the scope of this book.

13.3.5 Clean cities: air pollution reduction

To a degree, the observed significant improvements in urban air quality caused by the elimination of traditional air pollutants, such as soot, particles, and SO_2, in cities of high-income countries are a powerful illustration that cities act as innovation centers and hubs for environmental improvements that can lead to a sustainability transition path.

The first signs of progress in these traditional air pollutants are evident in countries of lower income as well, and are illustrated by the recent decline in the emissions of some traditional pollutants in Asian megacities. Nonetheless, an exceedingly high fraction of urban dwellers worldwide are still exposed to high levels of urban air pollution, especially TSPs with fine-particle emissions (PM10s) continuing their upward trend.

A wide portfolio of policy options is available, ranging from regulatory instruments such as mandated fuel choice ('smokeless zone' regulations), air-pollution standards, regulation of large point-source emissions and vehicle exhaust standards, to market-based instruments (or hybrid) approaches that incentivize technological change. An important feature of these regulatory or market-based approaches is *dynamic target setting* to reflect changing technology options and to counter the consequences of urban growth and potential consumer 'take-back' effects. The tested experiences in a diversity of settings from the United Kingdom through California to New Delhi provide valuable lessons for policy learning. However, an institutional locus on the exchange of policy lessons and capacity building (including pollution monitoring), particularly in small- and medium-sized cities in low-income countries, remains sorely lacking.

Air pollution is also the area of urban environmental policymaking where the most significant co-benefits of policies can be realized: improving access to clean cooking fuels, for example, improves human health and lowers traditional pollutant emissions, and also has (through reduced black carbon emissions) net global warming co-benefits. Similarly, improved energy efficiency and public transport options (e.g., New Delhi's transition to CNG buses) can also yield co-benefits on a variety of fronts (lower energy and transport costs for the poor, cleaner air, improved urban functionality) and should therefore be higher on the policy agenda than more single-purpose policy measures (e.g., renewable portfolio standards for urban electricity supply). Examples of the management of urban heat island effects also illustrate well the potential significant co-benefits between climate mitigation and adaptation measures.

Realization of the significant potential co-benefits, however, requires a holistic policy approach that integrates urban land-use, transport, and energy policies with the more traditional air-pollution policy frameworks. Cities in low-income countries, where the growth trends of urbanization and air pollutants are the most pronounced (to the extent that they will determine global trends), are also where institutional capacity gaps and information needs are largest. These must be improved to be able to reap the multiple co-benefits of more integrated urban policies. A renewed effort to improve measurement and monitoring as well as planning and modeling of urban environmental quality with a particular focus on urban energy and urban transport policy is urgently needed. This needs to be coupled with the development of institutional capacity for the design, implementation, and enforcement of policies and plans.

13.3.6 Policy leverages, priorities, and paradoxes

The highest impacts of urban policy decisions are in the areas where policies can affect local decision making and prevent or unblock spatial irreversibility or technological 'lock-in', or to steer away from critical thresholds.

Examples include preventing the further development of low-density, suburban housing and shopping malls, or promotion of the co-location of high energy supply and demand centers within a city that enable cogeneration and waste-heat recycling for heating and cooling purposes (e.g., in business districts). Buildings energy-performance standards are also a prime example of policy interventions that need to be implemented as early as possible to reap long-term benefits in terms of reduced energy use and improved urban environmental quality. New technologies, like smaller micro- and meso-grids are particularly attractive options that are also suitable for deregulated market environments. The literature on urban energy use, particularly with respect to transport, also identifies a critical 'threshold' of between 50 and 150 inhabitants/hectare below which public transport (or energy cogeneration) options become economically infeasible, which thus leads to over proportional increases in energy use (e.g., longer trips using private automobiles). To avoid such critical thresholds being crossed should be a high priority for urban administrations.

Given capital constraints, it is entirely unrealistic to expect 'grand' conceptual new urban 'eco-designs' to play any significant role in integrating some three billion additional urban dwellers to 2050 into the physical, economic, and social fabric of cities. (Building cities for these three billion new urban citizens along the Masdar (Abu Dhabi) model would require an investment to the tune of well above US$1,000 trillion, or some 20 years of current world GDP!) The role of such new, daring urban designs is less a template for development, but rather a 'learning laboratory' to develop and test approaches, especially low-cost options for sustainable urban growth in low-income countries and to retrofit and adapt existing urban structures and systems across the globe.

To address urban energy sustainability challenges will also require a new paradigm for drawing systems or ecosystems boundaries that extend the traditional place-based approach (e.g., based on administrative boundaries or ecosystems such as regional watersheds or air-quality districts). Sustainability criteria need to be defined on the basis of the functional interdependence among different systems, which are not necessarily in geographic proximity to each other. System analytical and extended LCA methods are increasingly available to address the question of the social, economic, and environmental sustainability of urban energy systems that almost exclusively rely on imports. However, clear methodological guidelines and strategies to overcome the formidable data challenges are needed, a responsibility that resides within the scientific community, but that requires support and a dedicated long-term approach for funding and capacity building.

A common characteristic of sustainable urban energy system options and policies is that they are usually systemic: for example, the integration of land-use and urban transport planning that extends beyond traditional administrative boundaries; the increasing integration of urban resource streams, including water, wastes, and energy, that can further both resource (e.g., heat) recovery and improve environmental performance; or the reconfiguration of urban energy systems toward a higher integration of supply and end-use (e.g., via micro- and meso-grids) that enable step changes in efficiency, for example, through cogeneration and energy cascading. This view of more integrated and more decentralized urban infrastructures also offers possibilities to improve the resilience and security of urban energy systems.

And yet this systemic perspective reveals a new kind of 'governance paradox'. Whereas the largest policy leverages are from systemic approaches and policy integration, these policies are also the most difficult to implement and require that policy fragmentation and uncoordinated, dispersed decision making be overcome. This governance paradox is compounded by weak institutional capacities, especially in small- to medium-sized cities that are the focus of projected urban growth, as well as by the legacies of market deregulation and privatization that have made integrated urban planning and energy, transport, and other infrastructural policy approaches more difficult to design and yet more difficult to implement.

However, there are good reasons for (cautionary) optimism. Urban areas will continue to act as innovation centers for experimentation and as diffusion nodes for the introduction of new systems and individual technological options by providing critical niche market sizes in the needed transition toward more sustainable urban energy systems. The task ahead is to leverage fully this innovation potential of cities and to scale up successful experiments into transformative changes in energy systems. Individual and collective learning, transfer of knowledge, and sharing experiences and information across cities and among stakeholders will, as always, be key, objectives to which this book hopes to contribute.

13.4 The urban future

The future is urban. In a world in which the scale of the human race's activities overwhelm almost every geophysical cycle it will also be a very different world. If the focus is to be urban then we need to recognize that the provision of energy is critical for the operation of the city. But urban energy policy brings with it some fundamental constraints. It has to be affordable to the city's poorest unskilled workers. That is a great pressure on the condition of the housing stock and those who charge rent for it. It has to be secure which is a great pressure on the city's form to deliver resilience in times of crisis. It has to be clean to ensure a better quality of life not just for the city's suburban elites but for all of its citizens. That places great pressure on the form of energy and the parsimony with which the city uses it. While in the past it seemed sufficient to face these problems with essentially upstream ('supply-side') policy interventions that no longer seems possible. Policy needs to address energy demand and that demand is *urban*.

Central policy, whatever it shortcomings is open to a national level of scrutiny. Local policy while more focused on real world delivery is also open to lobbying and local interests. This book hopes to show how analysis can help to deliver the advantages of local interventions without its disadvantages. The key is a sense of rigor in analysis. Such analysis provides sobering comparison with peers. Benchmarks are a vital part of this story. But no two cities even tens of miles apart are quite the same. Analysis determines what is unique in every context. The central policy maker provides an important framework but demand reduction without service reduction requires local policies appropriate to local circumstances. The most important step is to create the appropriate governance and analytic framework. Given frameworks fit for the task we really can energize our cities in the twenty-first century.

Notes

1   Important exceptions include the utilization of urban wastes and – where available – geothermal resources, both of which are characterized by high energy density.
2   Evidently, individual cities can forge ahead or fall behind the overall distributional pattern of aggregate uniform urban growth rates as outlined by the rank–size rule.
3   That is, a minimum density level over the entire settlement area that comprises residential zones of higher density with low-density green spaces.

References

Abiko, A., L. Cardoso, et al. (2007). "Basic costs of slum upgrading in Brazil." *Global Urban Development Magazine*. Washington, DC, USA, Global Urban Development. 3.

Adelekan, I. and A. Jerome (2006). "Dynamics of household energy consumption in a traditional African city, Ibadan." *Environmentalist* 26(2): 99–110.

Akbari H. and S. Bretz et al. (1997). "Peak power and cooling energy savings of high albedo roofs." *Energy and Buildings* 25:117–126.

ALA (2009). *State of the Air – 2008*. New York, NY, USA, American Lung Association (ALA).

Allen, P. M. (1997). *Cities and Regions as Self-Organizing Systems: Models of complexity*. London, UK, Routledge.

Alvarez, A. (2006). "Urban energy and poor women's enterprises in Salvador, Brazil." *Energia News* 9(1): 16–17.

Amaral, L. A. N. and J. M. Ottino (2004). "Complex networks: Augmenting the framework for the study of complex systems." *European Physical Journal B – Condensed Matter and Complex Systems* 38(2): 147–162.

Anderson, R. C. and R. D. Morgenstern (2009). "Marginal abatement cost estimates for non-CO_2 greenhouse gases: lessons from RECLAIM." *Climate Policy* 9: 40–55.

Andersson, E. (2006). "Urban landscapes and sustainable cities." *Ecology and Society* 11(1): 34.

Andrews, C. J. (2008). "Greenhouse gas emissions along the rural–urban gradient." *Journal of Environmental Planning and Management* 51(6): 847–870.

APHRC (2002). "Population and Health Dynamics in Nairobi's Informal Settlements." *Report of the Nairobi Cross-sectional Slums Survey (NCSS) 2000*. Nairobi, Kenya, African Population and Health Research Center (APHRC).

APPROTECH (2005). "Enabling Urban Poor Livelihood Policy Making: Understanding the Role of Energy Services in the Philippines." *National Workshop Reports*. Manila, Philippines, The Asian Alliance of Appropriate Technology Practitioners, Inc. (APPROTECH).

Arnold, J. E. M., G. Köhlin, et al. (2006). "Woodfuels, livelihoods, and policy interventions: Changing Perspectives." *World Development* 34(3): 596–611.

Arvesen, A., J. Liu, et al. (2010). "Energy cost of living and associated pollution for Beijing residents." *Journal of Industrial Ecology* 14(6): 890–901.

Balmer, M. (2007). "Household coal use in an urban township in South Africa." *Journal of Energy in Southern Africa* 18(3): 27–32.

Banham, R. (1969). *Architecture of the Well-tempered Environment*. Chicago, IL, USA, University of Chicago Press.

Banister, D. (1992). "Energy use, transport and settlement patterns." *Sustainable Development and Urban Form*. M. J. Breheny. London, UK, Pion Ltd.

Banister, D., S. Watson, et al. (1997). "Sustainable cities: transport, energy, and urban form." *Environment and Planning B: Planning and Design* 24(1): 125–143.

Barles, S. (2009). "Urban metabolism of Paris and its region." *Journal of Industrial Ecology* 13(6): 898–913.

Barter, P. A. (1999). "Transport and urban poverty in Asia. A brief introduction to the key issues." *Regional Development Dialogue* 20(1): 143–163.

Batty, M. (2005). *Cities and Complexity: Understanding Cities with Cellular Automata, Agent-Based Models and Fractals.* Cambridge, MA, USA, MIT Press.

Batty, M. (2008). "The size, scale, and shape of cities." *Science* 319(5864): 769–771.

Baumol, W. J. and W. E. Oates (1975). *The Theory of Environmental Policy: Externalities, Public Outlays, and the Quality of Life.* Englewood Cliffs, NJ, USA, Prentice-Hall.

Baynes, T. M. and X. Bai (2009). "Trajectories of change: Melbourne's population, urban development, energy supply and use from 1960–2006." Laxenburg, Austria, International Institute for Applied Systems Analysis (IIASA).

Baynes, T. M., M. Lenzen, et al. (2011). "Comparison of direct and indirect assessments of urban energy and implications for policy." *Energy Policy* 39(11): 7298–7309.

Beijing Government (2010). "Energy statistics online." Retrieved 10 December, 2010, from www.ebeijing.gov.cn/feature_2/Statistics/.

Bekker, B., A. Eberhard, et al. (2008). "South Africa's rapid electrification programme: Policy, institutional, planning, financing and technical innovations." *Energy Policy* 36(8): 3125–3137.

Berry, L., J. Brian, et al. (1958). "Alternate Explanations of Urban Rank–Size Relationships." *Annals of the Association of American Geographers* 48(1): 83–91.

Bettencourt, L. M. A., J. Lobo, et al. (2007). "Growth, innovation, scaling, and the pace of life in cities." *Proceedings of the National Academy of Sciences* 104(17): 7301–7306.

Bhan, G. (2009). "This is no longer the city I knew. Evictions, the urban poor and the right to the city in millennial Delhi." *Environment and Urbanization* 21(1):127–142.

Biello, D. (2008). "Eco-cities of the future." *Scientific American* (Earth 3.0): 6.

Block, A., K. Keuler, et al. (2004). "Impacts of anthropogenic heat on regional climate patterns." *Geophysical Research Letters* 31(12): L12211.

Bloomberg, M. R. (2007). plaNYC – a greener, greater New York. New York, City of New York: 158.

Boadi, K. O. and M. Kuitunen (2005). "Environment, wealth, inequality and the burden of disease in the Accra metropolitan area, Ghana." *International Journal of Environmental Health Research* 15(3): 193–206.

Boardman, B. (1993). *Fuel Poverty: From Cold Homes to Affordable Warmth.* London, UK, John Wiley & Sons.

Bocquier, P. (2005). "World Urbanization Prospects: an alternative to the UN model of projection compatible with the mobility transition theory." *Demographic Research* 12 (9): 197–236.

Böhm, R. (1998). "Urban bias in temperature time series – A case study for the city of Vienna, Austria." *Climatic Change* 38(1): 113–128.

Bravo, G., R. Kozulj, et al. (2008). "Energy access in urban and peri-urban Buenos Aires." *Energy for Sustainable Development* 12(4): 56–72.

BRE (2009). "SAP 2005. Building research establishment." Retrieved December 16, 2009, from www.bre.co.uk/sap2005.

Brehny, M. (1986). "Centrists, Decentrists and Compromisers: Views on the Future of Urban Form." *The Compact City. A Sustainable Urban Form?* M. Jenks, E. Burton and K. Williams. London, UK, Spon Press, Chapman and Hall.

Brög, W. and I. Ker (2009). "Myths (Mis)perceptions and reality in measuring voluntary behavioural changes." *Transport Survey Methods – Keeping Up With a Changing World.* P. Bonnel, M. Lee-Gosselin, J. Zmud and J. L. Madre. Bingley, UK, Emerald.

Brown, M. A., F. Southworth, et al. (2008). Shrinking the carbon footprint of metropolitan America. *Metropolitan Policy Program.* Washington, DC, Brookings.

Brownsword, R. A., P. D. Fleming, et al. (2005). "Sustainable cities: Modelling urban energy supply and demand." *Applied Energy* 82(2): 167–180.

Bruckner, T., R. Morrison, et al. (2003). "High-resolution modeling of energy-services supply systems using decco: overview and application to policy development." *Annals of Operations Research* 121: 151–180.

Buzar, S. (2006). "Estimating the extent of domestic energy deprivation through household expenditure surveys." *CEA Journal of Economics* 1(2): 6–19.

CAI (2009). *Clean Air Initiative for Asian Cities – Annual Report 2009.* Manila, Philippines, Clean Air Initiative (CAI).

CAI-Asia (2010). *Air Quality in Asia: Status and trends 2010 edition.* Clean Air Initiative-Asia (CAI-Asia), Manila, Philippines.

Carruthers, R., M. Dick, et al. (2005). "Affordability of public transport in developing countries." *Transport Paper 3* Washington, DC, USA, World Bank.

Cervero, R. and K. Kockelman (1997). "Travel demand and the 3Ds: Density, diversity, and design." *Transportation Research Part D: Transport and Environment* 2(3): 199–219.

Chambwera, M. and H. Folmer (2007). "Fuel switching in Harare: An almost ideal demand system approach." *Energy Policy* 35(4): 2538–2548.

Chandler, T. (1987). *Four Thousand Years of Urban Growth: A Historical Census.* New York, NY, USA, Edwin Mellen Press.

Chandler, T. and G. Fox (1974). *Three Thousand Years of Urban Growth.* New York, NY, USA, Academic Press.

Cherry, S. (2007). "How to build a green city." *IEEE Spectrum Online: Technology, Engineering, and Science News.* Retrieved 30 December, 2010, from spectrum. ieee.org/energy/environment/how-to-build-a-green-city/0.

Chertow, M. R. (2000). "The IPAT equation and its variants." *Journal of Industrial Ecology* 4(4): 13–29.

City of London (2007). *Rising to the Challenge – The City of London Corporation's Climate Adaptation Strategy.* London, UK, City of London Corporation: 95.

Cohen, C., M. Lenzen, et al. (2005). "Energy requirements of households in Brazil." *Energy Policy* 33(4): 555–562.

Cole, R. J. and P. C. Kernan (1996). "Life-cycle energy use in office buildings." *Building and Environment* 31(4): 307–317.

Cowan, B. (2008). "Identification and Demonstration of Selected Energy Best Practices for Low-Income Urban Communities in South Africa." *Alleviation of Poverty through the Provision of Local Energy Services (APPLES). Project No. EIE-04–168. Deliverable No. 17.* Cape Town, South Africa, Energy Research Centre, University of Cape Town and the Intelligent Energy Europe programme of the European Commission (EC).

Crovella, M. (2007). Personal communication on data published in Lakhina et al. 2003. Boston, MA, USA, Boston University.

Crutzen, P. J. (2004). "New directions: The growing urban heat and pollution island effect–impact on chemistry and climate." *Atmospheric Environment* 38(21): 3539–3540.

Cunningham-Sabot, E. and S. Fol (2009). "Shrinking cities in France and Great Britain: A silent process?" *The Future of Shrinking Cities: Problems, Patterns and Strategies of Urban Transformation in a Global Context*. K. Pallagst, J. Aber and I. Audirac. Berkeley, CA, USA, Center for Global Metropolitan Studies, Institute of Urban and Regional Development: 17–28.

Dayton, D., C. Goldman, et al. (1998). "The Energy Services Company (ESCO) Industry: Industry and Market Trends." *LBNL-41925*. Berkeley, CA, USA, Lawrence Berkeley National Laboratory.

Decker, E. H., A. J. Kerkhoff, et al. (2007). "Global patterns of city size distributions and their fundamental drivers." *PLoS ONE* 2(9): e934.

Decker, H., S. Elliott, et al. (2000). "Energy and material flow through the urban ecosystem." *Annual Review of Energy and the Environment* 25: 685–740.

Deevey, E. S. (1960). "The human populations." *Scientific American* 203(3): 194–204.

Devas, N. and D. Korboe (2000). "City governance and poverty: The case of Kumasi." *Environment and Urbanization* 12(1): 123–136.

Dey, C., C. Berger, et al. (2007). "An Australian environmental atlas: household environmental pressure from consumption." *Water, Wind, Art and Debate: how environmental concerns impact on disciplinary research*. G. Birch. Sydney, Australia, Sydney University Press: 280–315.

Dhakal, S. (2004). *Urban Energy Use and Greenhouse Gas Emissions in Asian Mega-cities: policies for a sustainable future*. H. Imura. Kitakyushu, Japan, Institute for Global Environmental strategies (IGES).

Dhakal, S. (2009). "Urban energy use and carbon emissions from cities in China and policy implications." *Energy Policy* 37(11): 4208–4219.

Dhakal, S. and K. Hanaki (2002). "Improvement of urban thermal environment by managing heat discharge sources and surface modification in Tokyo." *Energy and Buildings* 34(1): 13–23.

Dhakal, S., K. Hanaki, et al. (2003). "Estimation of heat discharges by residential buildings in Tokyo." *Energy Conversion and Management* 44(9): 1487–1499.

Dhakal, S. and L. Schipper (2005). "Transport and environment in Asian cities: Reshaping the issues and opportunities into a holistic framework." *International Review for Environmental Strategies* 5(2): 399–424.

Dhingra, Chavi, et al. (2008). "Access to clean energy services for the urban and peri-urban poor: a case-study of Delhi, India." *Energy for Sustainable Development* 12(4): 49–55.

DIALOG (2010). *Individuelle Motivation zum klimaschonenden Umgang mit Energie im Verkehr und im Haushalt* (Individual motivation for climate protecting energy use in transport and household). G. Sammer, R. Hössinger and J. Stark. Vienna, Austria, Klima & Energie Fonds, Institute for Transport Studies, University of Natural Resources and Applied Life Sciences.

Diamond, J. (2004). "The wealth of nations." *Nature* 429(6992): 616–617.

Doi, K., M. Kii, et al. (2008). "An integrated evaluation method of accessibility, quality of life, and social interaction." *Environment and Planning B: Planning and Design* 35(6): 1098–1116.

Doll, C. N. H. (2009). "Spatial Analysis of the World Bank's Global Air Pollution data Set." *IR-09–033*. Laxenburg, Austria, International Institute for Applied Systems Analysis (IIASA).

Doll, C. N. H. and S. Pachauri (2010). "Estimating rural populations without access to electricity in developing countries through night-time light satellite imagery." *Energy Policy* 38(10): 5661–5670.

Doxiadis, C. A. and J. G. Papaioannou (1974). "Ecumenopolis: The inevitable city of the future." *Environmental Conservation* 2(4): 315–317.

Drukker, C. (2000). "Economic consequences of electricity deregulation: A case study of San Diego gas & electric in a deregulated electricity market." *California Western Law Review* 361: 291.

Ejigie, D. A. (2008). "Household determinants of fuelwood choice in urban Ethiopia: A case study of Jimma Town." *Journal of Developing Areas* 41(1): 117–126.

Energie Wien (2009). "Energieflussbild Wien 2007." Retrieved 30 December, 2010, from www.wien.gv.at/wirtschaft/eu-strategie/energie/zahlen/energiever brauch.html.

ESMAP (2006). "Pakistan: Household Use of Commercial Energy." *Report No. 320/06*. M. Kojima. Washington, DC, USA, Energy Sector Management and Assistance Programme (ESMAP) of the World Bank.

ESMAP (2007). "Meeting the Energy Needs of the Urban Poor: Lessons from Electrification Practitioners." *Technical Paper No. 118/07*. Washington, DC, USA, Energy Sector Management Assistance Program (ESMAP) of the World Bank.

ESMAP and UNDP (2003). "Household Fuel Use and Fuel Switching in Guatemala." *Report No. 27274*. Washington, DC, USA, Energy Sector Management Assistance Programme (ESMAP) of the World Bank and the United Nations Development Programme (UNDP).

ESMAP and UNDP (2005). "Power Sector Reform in Africa: Assessing Impact on Poor People." *Report No. 306/05*. Washington, DC, USA, Energy Sector Management Assistance Programme (ESMAP) of the World Bank and United Nations Development Programme (UNDP).

Eurostat (1988). *Useful Energy Balances*. Luxembourg, Statistical Office of the European Communities (Eurostat).

Eurostat (2008). *Gross domestic product (GDP) at current market prices at NUTS level 3*, Eurostat.

Ewing, R. and R. Cervero (2001). "Travel and the Built Environment: A Synthesis." *Transportation Research Record: Journal of the Transportation Research Board* 1780(1): 87–114.

Faist, M., R. Frischknecht, et al. (2003). *Métabolisme des Activités Économiques du Canton de Genève – Phase 1*. Geneva, Switzerland, Groupe de Travail Interdépartemental Ecosite.

Fall, A., S. Sécou, et al. (2008). "Modern energy access in peri-urban areas of West Africa: the case of Dakar, Senegal." *Energy for Sustainable Development* 12(4): 22–37.

Fisher, J. C. and R. H. Pry (1971). "A simple substitution model of technological change." *Technological Forecasting & Social Change* 3: 75–88.

Fisk, D. (2008). "What are the risk-related barriers to, and opportunities for, innovation from a business perspective in the UK, in the context of energy management in the built environment?" *Energy Policy* 36(12): 4615–4617.

Fisk, D. (2010). Exergy analyses and Sankey diagrams for the cities of London and Malmö. Personal Communication. London, UK, Imperial College London.

Fisk, D. J. and J. Kerhervéa (2006). "Complexity as a cause of unsustainability "*Ecological Complexity* 3(4): 336–343.

Flanner, M. G. (2009). "Integrating anthropogenic heat flux with global climate models." *Geophysical Research Letters* 36(2): L02801.

Forrester, J. W. (1969). *Urban Dynamics*. Waltham, MA, Pegasus Communications, Inc.

Friends of the Environment (2005). "Enabling urban poor livelihoods policy making: Understanding the role of energy services: Country study Nigeria." UK Department for International Development. http://www.dfid.gov.uk/r4d/PDF/Outputs/Energy/R8348-Nigeria.pdf.

Fujita, M., P. Krugman, et al. (1999). *The Spatial Economy: cities, regions and international trade*. Cambridge, MA, USA, MIT Press.

Gangopadhyay, S., B. Ramaswami, et al. (2005). "Reducing subsidies on household fuels in India: How will it affect the poor?" *Energy Policy* 33(18): 2326–2336.

Gasson, B. (2007). *Aspects of stocks and flows in the built environment of Cape Town*. Workshop on Analysing Stocks and Flows of the Built Urban Environment, Trondheim, Norway, Norwegian University of Science and Technology (NTNU).

Gilli, P. V., N. Nakicenovic, et al. (1995). *First and Second Law Efficiencies of the Global and Regional Energy Systems* PS 3.1.16, 16th congress of the World Energy Congress (WEC), Tokyo, Japan.

Gilli, P. V., N. Nakicenovic, et al. (1996). "First- and second-law efficiencies of the global and regional energy systems." *RR-96–2*. Laxenburg, Austria, International Institute for Applied Systems Analysis (IIASA).

Giradin, L. and D. Favrat (2010). "Bilan Énergétique/Exergétique de Genève pour 2005." Unpublished ms. Lausanne, Switzerland, École Polytechnique Fédérale de Lausanne (EPL).

GLA (2004). "Green light to clean power: The mayor's energy strategy." From: www.london.gov.uk/mayor/strategies/energy/index.jsp.

Goldemberg, J. (1991). ""Leap-frogging": A new energy policy for developing countries." *WEC Journal* (December): 27–30.

Goyal, P. (2003). "Present scenario of air quality in Delhi: A case study of CNG implementation." *Atmospheric Environment* 37(38): 5423–5431.

Grubler, A. (1994). "Technology." *Changes in Land Use and Land Cover: A Global Perspective.* W. B. Meyer and B. L. Turner II. Cambridge, UK, Cambridge University Press: 287–328.

Grubler, A. (2004). "Transitions in Energy Use." *Encyclopedia of Energy.* C. J. Cleveland. Amsterdam, the Netherlands, Elsevier Science. 6: 163–177.

Grubler, A., B. O'Neill, et al. (2007). "Regional, national, and spatially explicit scenarios of demographic and economic change based on SRES." *Technological Forecasting and Social Change* 74(7): 980–1029.

Guimerà, R., S. Mossa, et al. (2005). "The worldwide air transportation network: Anomalous centrality, community structure, and cities' global roles." *Proceedings of the National Academy of Sciences of the United States of America* 102(22): 7794–7799.

Gutschner, M., S. Nowak, et al. (2001). "Potential for Building Integrated Photovoltaics." *Report IEA PVPS T7–4–2001 (Summary)*. Paris, France, International Energy Agency (IEA) of the Organisation for Economic Co-operation and Development (OECD).

Haberl, H. (2001). "The energetic metabolism of societies, Part 1: accounting concepts." *Journal of Industrial Ecology* 5(1): 11–33.

Hall, P. (2002). *Cities of Tomorrow*. Oxford, UK, Wiley-Blackwell.

Hardoy, J. E., D. Mitlin, et al. (2001). *Environmental Problems in an Urbanizing World*. London, UK, Earthscan.

Harms, H. (1997). "To live in the city centre: housing and tenants in central neighbourhoods of Latin American cities." *Environment and Urbanization* 9(2):191–212.

Harrison, D. (2004). Ex post evaluation of the RECLAIM emissions trading programmes for the Los Angeles air basin. *Tradable Permits: Policy Evaluation, Design and Reform*. Paris, France, Organisation for Economic Co-operation and Development (OECD): 192.

HEI (2010a). *Outdoor Air Pollution and Health in the Developing Countries of Asia: A Comprehensive Review*, Health Effects Institute, Special Report 18, Boston, MA, USA.

HEI (2010b). "Public Health and Air Pollution in Asia (PAPA): Coordinated Studies of Short-Term Exposure to Air Pollution and Daily Mortality in Four Cities," Health Effects Institute, Research Report No 154, Boston, MA, USA.

Hersey, J., N. Lazarus, et al. (2009). "Capital Consumption: The Transition to Sustainable Consumption and Production in London." London, UK, BioRegional and London Sustainable Development Commission: 80.

Hillier, B. (1999). *Space is the Machine – A Configurational Theory of Architecture*. Cambridge, UK, Cambridge University Press.

Hillman, T. and A. Ramaswami (2010). "Greenhouse gas emission footprints and energy use benchmarks for eight U.S. cities." *Environmental Science & Technology* 44(6): 1902–1910.

Holden, E. and I. T. Norland (2005). "Three challenges for the compact city as a sustainable urban form: Household consumption of energy and transport in eight residential areas in the greater Oslo region." *Urban Studies* 42(12): 2145–2166.

Hu, M., H. Bergsdal, et al. (2010). "Dynamics of urban and rural housing stocks in China." *Building Research & Information* 38(3): 301–317.

Hu, P. S. and T. R. Reuscher (2004). *Summary of Travel Trends – 2001 National Household Travel Survey*. Washington, DC, USA, Federal Highway Administration, United States Department of Transportation.

Huq, A. T., M. Zahurul, et al. (1996). Transport and the Urban Poor. *The Urban Poor in Bangladesh*. N. Islam. Dhaka, Bangladesh, Centre for Urban Studies: 123.

IAUC and WMO (2006). *Preprints: Sixth International Conference on Urban Climate*. Göteborg, Sweden, International Association for Urban Climate (IAUC), the World Meteorological Organisation (WMO) and Göteborg University.

ICARO (1999). "Increase of car Occupancy through Innovative Measures and Technical Instruments." G. Sammer. Vienna, Austria, European Commission under the 4th Framework Program, ICARO Consortium.

Ichinose, T. (2008). "Anthropogenic Heat and Urban Heat Islands: A Feedback System." *Newsletter on Urban Heat Island Counter Measures*. Tokyo, Japan, Committee on the Global Environment, Architectural Institute of Japan. 5.

ICLEI (2009). "Harmonized Emissions Analysis Tool." Oakland, CA, USA, International Council of Local Environment Initiatives (ICLEI).

IEA (2002). *World Energy Outlook 2002*. Paris, France, International Energy Agency (IEA) of the Organisation of Economic Co-operation and Development (OECD).

IEA (2005). *Saving Oil in a Hurry*. Paris, France, International Energy Agency (IEA) of the Organisation for Economic Co-operation and Development (OECD).

IEA (2008). *World Energy Outlook 2008*. Paris, France, International Energy Agency (IEA) of the Organisation for Economic Co-operation and Development (OECD).

IEA (2010a). *Energy Balances of Non-OECD Countries*. Paris, France, International Energy Agency (IEA) of the Organisation for Economic Co-operation and Development (OECD).

IEA (2010b). *Energy Balances of OCED Countries*. Paris, France, International Energy Agency (IEA) of the Organisation for Economic Co-operation and Development (OECD).

IIASA (2010). IPCC Reference Concentration Pathways (RCP) Scenario Database, International Institute for Applied Systems Analysis (IIASA), Vienna, Austria.

Ijiri, Y. and H. A. Simon (1975). "Some distributions associated with Bose Einstein statistics." *Proceedings of the National Academy of Sciences of the United States of America* 72(5): 1654–1657.

IMF (2010). International Financial Statistics 2010. *CD-ROM Edition*. Washington, DC, USA, International Monetary Fund (IMF).

IPCC (2007). *Climate Change 2007. The Physical Science Basis. Working Group I Contribution to the Fourth Assessment Report.* Cambridge, UK, Intergovernmental Panel on Climate Change (IPCC), Cambridge University Press.

Irwin, E., K. Bell, et al. (2009). "The economics of urban–rural space." *Annual Review of Resource Economics* 1: 435–459.

Jabareen, Y. R. (2006). "Sustainable Urban Forms." *Journal of Planning Education and Research* 26(1): 38–52.

Jabeen, Huraera, et al. (2010). "Built-in resilience: learning from grassroots coping strategies to climate variability." *Environment and Urbanization* 22(2): 415–431.

Jalihal, S. A. and T. S. Reddy (2006). "Assessment of the impact of improvement measures on air quality: Case study of Delhi." *Journal of Transportation Engineering* 132(6): 482–488.

Jenks, M., E. Burton, et al. (1996). *The Compact City. A sustainable urban form?* London, UK, Spon Press, Chapman and Hall.

Joly, I. (2004). "The link between travel time budget and speed: A key relationship for urban space–time dynamics." Proceedings European Transport Conference 2004, Strasbourg, France.

Kalnay, E. and M. Cai (2003). "Impact of urbanization and land-use change on climate." *Nature* 423(6939): 528–531.

Kanda, M. (2006). "Progress in the scale modeling of urban climate: Review." *Theoretical and Applied Climatology* 84(1): 23–33.

Kandlikar, M. (2007). "Air pollution at a hotspot location in Delhi: Detecting trends, seasonal cycles and oscillations." *Atmospheric Environment* 41(28): 5934–5947.

Karekezi, S., J. Kimani, et al. (2008). "Energy access among the urban poor in Kenya." *Energy for Sustainable Development* 12(4): 38–48.

Kataoka, K., F. Matsumoto, et al. (2009). "Urban warming trends in several large Asian cities over the last 100 years." *Science of the Total Environment* 407(9): 3112–3119.

Kathuria, V. (2004). "Impact of CNG on Vehicular Pollution in Delhi: A Note." *Transportation Research Part D: Transport and Environment* 9(5): 409–417.

Kebede, B., A. Bekele, et al. (2002). "Can the urban poor afford modern energy? The case of Ethiopia." *Energy Policy* 30(11–12): 1029–1045.

Keirstead, J., N. Samsatli, et al. (2009). "Syncity: An Integrated Tool Kit for Urban Energy Systems Modelling." *Fifth Urban Research Symposium.* Marseilles, France, World Bank: 19.

Kelly, J., P. Jones, et al. (2004). "Volume 1: Concepts and Tools;" "Volume 2: Fac Sheets." *Successful Transport Decision-Making: A Project Management and Stakeholder Engagement Handbook.* Brussels, Belgium, European Commission under the 5th Framework Program, GUIDEMAPS Consortium.

Kennedy, C., J. Steinberger, et al. (2009). "Greenhouse gas emissions from global cities." *Environmental Science & Technology* 43(19): 7297–7302.

Kennedy, C., J. Steinberger, et al. (2010). "Methodology for inventorying greenhouse gas emissions from global cities." *Energy Policy* 38(9): 4828–4837.

Kenworthy, J. R. (2006). "The eco-city: Ten key transport and planning dimensions for sustainable city development." *Environment and Urbanization* 18(1): 67–85.

Kenworthy, J. and F. B. Laube (2001). Millennium Database for Sustainable Transport. Brussels, Belgium, International Union of Public Transport (UITP).

Kenworthy, J. R. and F.B Laube, et al. (1999). *An International Sourcebook of Automobile Dependence in Cities 1960–1990*. University Press of Colorado, Boulder, CO, USA.

Kenworthy, J. R. and P. W. G. Newman (1990). "Cities and transport energy – lessons from a global survey." *Ekistics – the Problems and Science of Human Settlements* 57(344–45): 258–268.

Kikegawa, Y., Y. Genchi, et al. (2003). "Development of a numerical simulation system toward comprehensive assessments of urban warming countermeasures including their impacts upon the urban buildings' energy-demands." *Applied Energy* 76(4): 449–466.

Kim, K. S. and N. Gallent (1998). "Regulating industrial growth in the South Korean capital region." *Cities* 15(1): 1–11.

Kinoshita, T., E. Kato, et al. (2008). "Investigating the rank–size relationship of urban areas using land cover maps." *Geophys. Res. Lett.* 35(17): L17405.

Kluy, A. (2010). "Wenn das Öl zu Ende geht." *Die Welt*: 25.

Klysik, K. and K. Fortuniak (1999). "Temporal and spatial characteristics of the urban heat island of Lódz, Poland." *Atmospheric Environment* 33(24–25): 3885–3895.

Kölbl, R. and D. Helbing (2003). "Energy laws in human travel behaviour." *New Journal of Physics* 5: 48.41–48.12.

Krugman, P. (1991). "Increasing returns and economic geography." *Journal of Political Economy* 99(3): 483–499.

Krugman, P. (1996). "Confronting the mystery of urban hierarchy." *Journal of the Japanese and international economics* 10: 399–418.

Kühnert, C., D. Helbing, et al. (2006). "Scaling laws in urban supply networks." *Physica A: Statistical Mechanics and its Applications* 363(1): 96–103.

Kulkarni, A., G. Sant, et al. (1994). "Urbanization in search of energy in three Indian cities." *Energy* 19(5): 549–560.

Kyokutamba, J. (2004). "Uganda." *Energy services for the urban poor in Africa: Issues and policy implications*. B. Kebede and I. Dube. London, UK, Zed Books: 231–278.

Larivière, I. and G. Lafrance (1999). "Modelling the electricity consumption of cities: Effect of urban density." *Energy Economics* 21(1): 53–66.

Larson, C. (2009). "China's grand plans for eco-cities now lie abandoned." *Yale Environment 360*. New Haven, CT, USA, Yale School of Forestry & Environmental Studies.

Leach, G. and R. Mearns (1989). *Beyond the Woodfuel Crisis – People, land and trees in Africa*. London, UK, Earthscan

Ledent, J. (1980). "Comparative Dynamics of three demographic models of urbanization." Laxenburg, Austria, IIASA. RR-80–1.

Ledent, J. (1982). "Rural–urban migration, urbanization, and economic development." *Economic Development and Cultural Change* 30(3): 507–538.

Ledent, J. (1986). "A model of urbanization with nonlinear migration flows." *International Regional Science Review* 10(3): 221–242.

Lenton, T. M., H. Held, et al. (2008). "Tipping elements in the Earth's climate system." *Proceedings of the National Academy of Sciences* 105(6): 1786–1793.

Lenzen, M., C. Dey, et al. (2004). "Energy requirements of Sydney households." *Ecological Economics* 49(3): 375–399.

Lenzen, M., M. Wier, et al. (2006). "A comparative multivariate analysis of household energy requirements in Australia, Brazil, Denmark, India and Japan." *Energy* 31(2–3): 181–207.

Levinson, D. M. and A. Kumar (1997). "Density and the journey to work." *Growth and Change* 28(2): 147–172.

Lin, J. J. and C. M. Feng (2003). "A bi-level programming model for the land use–network design problem." *Annals of Regional Science* 37(1): 93–105.

Link, Ch., U. Raich, G. Sammer and J. Stark (2011a). "Nutzernachfrage in definiertenSzenarien der SMART-ELECRIC-MOBILITY (Arbeitspaket 6) - Speichereinsatzfür regenerative elektrischeMobilität und Netzstabilität" (User demand of defined scenarios of SMART-ELECRIC-MOBILITY – Workpackage 6); Research Project of the Climate and Energy Fund Austria, Institute for Transport Studies at the University for Natural Resources and Life Sciences Vienna.

Link, Ch., U. Raich, G. Sammer and J. Stark (2011b). "Bewertung von Elektromobilitätsszenarien, Projekt SMART-ELECRIC-MOBILITY (Arbeitspaket 7) – Speichereinsatzfür regenerative elektrischeMobilität und Netzstabilität" (Assessment of electric mobility scenarios, projekct SMART-ELECRIC-MOBILITY – Workpackage 7); Research Project of the Climate and Energy Fund Austria, Institute for Transport Studies at the University for Natural Resources and Life Sciences Vienna.

Liu, J., G. C. Daily, et al. (2003). "Effects of household dynamics on resource consumption and biodiversity." *Nature* 421(6922): 530–533.

Longhurst, J. W. S., J. G. Irwin, et al. (2009). "The development of effects-based air quality management regimes." *Atmospheric Environment* 43(1): 64–78.

Ludwig, D., M. Klärle, et al. (2008). "Automatisierte Standortanalyse für die Solarnutzung auf Dachflächen über hochaufgelöste Laserscanningdaten." *Angewandte Geoinformatik 2008 – Beiträge zum 20. AGIT-Symposium Salzburg.* J. Strobl, T. Blaschke and J. Grieser. Heidelberg, Germany, Wichmann: 466–475.

Ma, K.-R. and D. Banister (2006). "Excess commuting: A critical review." *Transport Reviews* 26(6): 749–767.

McGranahan, G. and P. J. Marcotullio, Eds. (2007). *Urban Transitions and the Spatial Displacement of Environmental Burdens. Scaling Urban Environmental Challenges: From the local to global and back.* London, UK, Earthscan.

McGranahan, G., P. Jacobi, et al. (2001). *Citizens at Risk: From urban sanitation to sustainable cities.* London, UK, Earthscan.

Maibach, M., C. Schreyer, et al. (2008). "Handbook on Estimation of External Costs in the Transport Sector." *Internalisation Measures and Policies for All External Cost of Transport (IMPACT).* Delft, the Netherlands, CE Delft.

Marchetti, C. (1994). "Anthropological invariants in travel behavior." *Technological Forecasting and Social Change* 47(1): 75–88.

Masera, O. R. and G. S. Dutt (1991). "A thermodynamic analysis of energy needs: A case study in A Mexican village." *Energy* 16(4): 763–769.

Matthews, E., C. Amann, et al. (2000). *The Weight of Nations: Material Outflows from Industrial Economies, Material Outflows from Industrial Economics.* Washington, D.C, World Resources Institute.

Mayor of London (2004). *Greenlight to Clean Power: The mayor's energy strategy.* London, UK, Greater London Authority.

Meikle, S. and P. North (2005). "A study of the impact of energy use on poor women and girls' livelihoods in Arusha, Tanzania." *R8321.* London, UK, UK Department for International Development (DfID).

Meschik, M. (2008). *Planungshandbuch Radverkehr (Planning Handbook of Bicycle Traffic).* Vienna, Austria and New York, NY, USA, Springer.

Mestl, H. E. S., K. Aunan, et al. (2007). "Urban and rural exposure to indoor air pollution from domestic biomass and coal burning across China." *Science of the Total Environment* 377(1): 12–26.

Metz, D. (2008). *The Limits to Travel: How far will you go?* London, UK, Earthscan.

Minx, J. C., G. Feng, et al. (2011). "The carbon footprint of cities." Paper prepared for submission to the *Journal of Industrial Ecology*.

Minx, J. C., T. Wiedman, et al. (2009). "Input-Output analysis and carbon footprint: An overview of applications." *Economic Systems Research* 21(3):187–216.

Mitlin, D. (1997). "Tenants: addressing needs, increasing options." *Environment and Urbanization* 9(2):17–212.

MOE (2005). *2004 Status of Air Pollution*. Tokyo, Ministry of the Environment (MOE), Government of Japan.

Mwampamba, T. H. (2007). "Has the fuelwood crisis returned? Urban charcoal consumption in Tanzania and its implications to present and future forest availability." *Energy Policy* 35(8): 4221–4234.

Mwangi, I. K. (1997). "The nature of rental housing in Kenya." *Environment and Urbanization* 9(2):141–159.

Nakicenovic, N., A. Grübler, et al. (1998). *Global Energy Perspectives*. Cambridge, UK, Cambridge University Press.

Nakicenovic, N., A. Gruebler, et al. (1990). "Technological progress, structural change and efficient energy use: Trends worldwide and in Austria." Retrieved 20 May, 2010.

National Research Council (2003). *Cities Transformed: Demographic change and its implications in the developing world*. Washington, DC, USA, National Academies Press.

Newcombe, K., J. D. Kalma, et al. (1978). "The metabolism of a city: The case of Hong Kong." *Ambio* 7(1): 3–15.

Newman, P. and J. Kenworthy (1991). *Cities and Automobile Dependence: An international source book*. Aldershot, UK, Avebury.

Newman, P. W. G. and J. R. Kenworthy (1989). "Gasoline consumption and cities – A comparison of U.S. cities with a global survey." *Journal of the American Planning Association* 55(1): 24–37.

Newman, P. W. G. and J. R. Kenworthy (1999). *Sustainability and Cities: Overcoming automobile dependence*. Washington, DC, USA, Island Press.

Newton, P., S. Tucker, et al. (2000). "Housing form, energy use and greenhouse gas emissions." *Achieving Sustainable Urban Form*. K. Williams, E. Burton and M. Jenks. London, UK, Routledge: 74–84.

Nicholls, R. J., S. Hanson, et al. (2008). "Ranking Port Cities with High Exposure and Vulnerability to Climate Extremes." *OECD Environment Working Papers No. 1*. Paris, France, Organisation for Economic Co-operation and Development (OECD): 63.

Nijkamp, P. and S. A. Rienstra (1996). Sustainable transport in a compact city. *The Compact City. A sustainable urban form?* M. E. Jenks, E. Burton and K. Williams. London, UK, Spon Press, Chapman and Hall: 190–199.

NOAA (2008). Version 4 DMSP-OLS Nighttime Lights Series, National Geophysical Data Center, National Oceanic and Atmospheric Administration (NOAA).

Norman, J., H. L. MacLean, et al. (2006). "Comparing high and low residential density: Life-cycle analysis of energy use and greenhouse gas emissions." *Journal of Urban Planning and Development* 132(1): 10–21.

Normile, D. (2008). "China's living laboratory in urbanization." *Science* 319(5864): 740–743.

Oke, T. R. (1987). *Boundary Layer Climates*. New York, NY, USA, Routledge.

Oke, T. R. (2006). "Initial Guidance to Obtain Representative Meteorological Observations at Urban Sites." *Instruments and Observing Methods Report No. 81.* Geneva, Switzerland, World Meteorological Organisation.

Oleson, K. W., G. B. Bonan, et al. (2010). "An examination of urban heat island characteristics in a global climate model." *International Journal of Climatology:* n/a-n/a.

O'Neill, B., D. Balk, et al. (2001). "A guide to global population projections." *Demographic Research* 4: 203–288.

O'Neill, B. and S. Scherbov (2006). "Interpreting UN Urbanization Projections Using a Multi-state Model." Laxenburg, Austria, International Institute for Applied Systems Analysis. IR-06-012 i–17.

O'Neill, B. C., M. Dalton, et al. (2010). "Global demographic trends and future carbon emissions." *Proceedings of the National Academy of Sciences* 107(41): 17521–17526.

Ouedraogo, B. (2006). "Household energy preferences for cooking in urban Ouagadougou, Burkina Faso." *Energy Policy* 34(18): 3787–3795.

ÖVG (2009). *Handbuch Öffentlicher Verkehr, Schwerpunkt Österreich.* Vienna, Austria, Bohmann Druck und Verlag.

Pacala, S. and R. Socolow (2004). "Stabilization wedges: Solving the climate problem for the next 50 years with current technologies." *Science* 305(5686): 968–972.

Pachauri, S. (2004). "An analysis of cross-sectional variations in total household energy requirements in India using micro survey data." *Energy Policy* 32(15): 1723–1735.

Pachauri, S., et al. (2007). *An Energy Analysis of Household Consumption: Changing patterns of direct and indirect use in India.* Berlin, Germany, Springer-Verlag.

Pachauri, S. and L. Jiang (2008). "The household energy transition in India and China." *Energy Policy* 36(11): 4022–4035.

Pachauri, S. and D. Spreng (2002). "Direct and indirect energy requirements of households in India." *Energy Policy* 30(6): 511–523.

Pachauri, S., A. Mueller, et al. (2004). "On measuring energy poverty in Indian households." *World Development* 32(12): 2083–2104.

Padam, S. and S. K. Singh (2001). *Urbanization and Urban Transport in India: The sketch for a policy.* Pune, India, Central Institute of Road Transport.

Parshall, L., K. Gurney, et al. (2010). "Modeling energy consumption and CO_2 emissions at the urban scale: Methodological challenges and insights from the United States." *Energy Policy* 38(9): 4765–4782.

Patterson, W. (2009). *Keeping the Lights On.* London, UK, Earthscan.

Pearce, F. (2010). "The Population Crash." *Guardian.* London, UK.

Pedersen, P. D. (2007). "Human Development Report 2007/2008 – Mitigation Country Studies: Japan." *Occasional Paper No. 2007/56.* New York, NY, USA, Human Development Report Office, United Nations Development Programme (UNDP): 27.

Pelling, M. and B. Wisner, eds. (2009). *Disaster Risk Reduction: Cases from Urban Africa.* London, UK, Earthscan.

Peña, S. (2005). "Recent developments in urban marginality along Mexico's northern border." *Habit International* 29(2): 285–301.

Permana, A. S., R. Perera, et al. (2008). "Understanding energy consumption pattern of households in different urban development forms: A comparative study in Bandung City, Indonesia." *Energy Policy* 36(11): 4287–4297.

Peters, G., T. Briceno, et al. (2004). "Pollution Embodied in Norwegian Consumption." *Working Paper No. 6.* Trondheim, Norway, Industrial Ecology Programme, Norwegian University of Science and Technology (NTNU).

Peters, G. P. and E. G. Hertwich (2008). "CO_2 embodied in international trade with implications for global climate policy." *Environmental Science & Technology* 42(5): 1401–1407.

Pickett, S. T. A., M. L. Cadenasso, et al. (2008). "Urban Ecological Systems: Linking Terrestrial Ecological, Physical, and Socioeconomic Components of Metropolitan Areas." *Urban Ecology*. J. M. Marzluff, E. Shulenberger, W. Endlicher et al. New York, NY, USA, Springer US: 99–122.

Pischinger, R., S. Hausberger, et al. (1997). "Volkswirtschaftliche Kosten-Wirksamkeitsanalyse von Maßnahmen zur Reduktion der CO_2-Emissionen in Österreich." Graz, Vienna and Linz, Austria, Bundesministeriums für Umwelt, Jugend und Familie and the Akademie für Umwelt und Energie.

Pohekar, S. D., D. Kumar, et al. (2005). "Dissemination of cooking energy alternatives in India – A review." *Renewable and Sustainable Energy Reviews* 9(4): 379–393.

PriceWaterhouseCoopers (2007). *UK Economic Outlook 2007*. London, PriceWaterhouseCoopers LLP: 40.

Proops, J. L. R. (1977). "Input–output analysis and energy intensities: A comparison of some methodologies." *Applied Mathematical Modelling* 1(March): 181–186.

Ramaswami, A., A. Chavez, et al. (2011). "Two approaches to greenhouse gas emissions foot-printing at the city scale." *Environmental Science & Technology* 45(10): 4205–4206.

Ramaswami, A., T. Hillman, et al. (2008). "A demand-centered, hybrid life-cycle methodology for city-scale greenhouse gas inventories." *Environmental Science & Technology* 42(17): 6455–6461.

Ratti, C., N. Baker, et al. (2005). "Energy Consumption and Urban Texture." *Energy and Buildings* 37(7): 762–776.

Ravindra, K., E. Wauters, et al. (2006). "Assessment of air quality after the implementation of compressed natural gas (CNG) as fuel in public transport in Delhi, India." *Environmental Monitoring and Assessment* 115(1–3): 405–417.

Rees, W. and M. Wackernagel (1996). "Urban ecological footprints: why cities cannot be sustainable – And why they are a key to sustainability." *Environmental Impact Assessment Review* 16(4–6): 223–248.

Riahi, K., F. Dentener, et al. (in press). "Energy Pathways for Sustainable Development." *Global Energy Assessment: Toward a sustainable future*, IIASA, Laxenburg, Austria and Cambridge University Press, Cambridge, UK and New York, NY, USA.

Rickaby, P. A. (1991). "Energy and urban development in an archetypal English town." *Environment and Planning B: Planning and Design* 18(2): 153–175.

RITA (2009). "Integrated corridor management." Retrieved 30 December, 2009, from www.its.dot.gov/icms/index.htm.

Rogers, A. (1982). "Sources of urban population growth and urbanization, 1950–2000: A demographic accounting." *Economic Development and Cultural Change* 30(3): 483–506.

Romero Lankao, P., H. López Villafranco, et al. (2005). *Can Cities Reduce Global Warming? Urban development and the carbon cycle in Latin America*. Mexico City, Mexico, IAI, UAM-X, IHDP, GCP.

Rosen, M. A. (1992). "Evaluation of energy utilization efficiency in Canada using energy and exergy analyses." *Energy* 17(4): 339–350.

Rosenfeld, A. H., H. Akbari, et al. (1995). "Mitigation of urban heat islands: Materials, utility programs, updates." *Energy and Buildings* 22(3): 255–265.

Sabry, S. (2009). *Poverty Lines in Greater Cairo: Underestimating and misrepresenting poverty*. London, UK, International Institute for Environment and Development (IIED).

Sahakian, M. and J. K. Steinberger (2010). "Energy reduction through a deeper understanding of household consumption: Staying cool in metro Manila." *Journal of Industrial Ecology* 15(1): 31–48.

Salat, S. and C. Guesne (2008). "Energy and carbon efficiency of urban morphologies. The case of Paris." Paris, France, Urban Morphologies Laboratory, CSTB (French Scientific Centre for Building Research and ENSMP (École Nationale Supérieure des Mines de Paris).

Salat, S. and A. Morterol (2006). "Factor 20: A multiplying method for dividing by 20 the carbon energy footprint of cities: the urban morphology factor." Paris, France, Urban Morphologies Laboratory, CSTB (French Scientific Centre for Building Research) and ENSMP (École Nationale Supérieure des Mines de Paris).

Salon, D. and S. Gulyani (2010). "Mobility, poverty, and gender: Travel 'choices' of slum residents in Nairobi, Kenya." *Transport Reviews* 30(5): 641–657.

Sammer, G. (2008). *Economic Cost-Effectiveness of TDM-Measures regarding their Environmental Impact*. Vienna-Semmering.

Sammer, G. (2009a). *Konjunkturprogramme für Verkehr – Chancen für neue Wege im Verkehr?* 17. Bad Kreuznacher Verkehrssymposium Kohle für den Verkehr! Auswirkungen der Wirtschaftskrise auf die Mobilität, Bad Kreuznach, Germany.

Sammer, G. (2009b). "Non-Negligible Side Effects of Traffic Management." *Travel Demand Management and Road User Pricing: Success, failure and feasibility*. W. Saleh and G. Sammer. Surrey, UK, Ashgate: 268.

Sammer, G., W. J. Berger, et al. (2009). "Schriftliche Unterlagen Vekehrsplanung und Mobility." *Lehrveranstaltung 856 102*. Vienna, USA, Institut für Vekehrswesen, Universität für Bodenkultur.

Sammer, G., R. Klementschitz, et al. (2007). *A parking management scheme for private car parks – A promising approach to mitigate congestion on urban roads?* 23rd World Road Congress, Paris, France.

Sammer, G., J. Stark, et al. (2005). "Instrumente zur Steuerung des Stellplatzangebotes für den Zielverkehr." *Final Report, Parts 1 and 2*. Vienna, Austria, In-Stella Consortium, University of Natural Resources and Applied Life Sciences.

Sartori, I. and A. G. Hestnes (2007). "Energy use in the life cycle of conventional and low-energy buildings: A review article." *Energy and Buildings* 39(3): 249–257.

Satterthwaite, D. (2009). "The implications of population growth and urbanization for climate change." *Environment and Urbanization* 21(2): 545–567.

Satterthwaite, D. (2010). "Urban myths and the mis-use of data that underpin them." *Urbanization and Development: Multidisciplinary perspectives*. J. Beall, B. Guha-Khasnobis and R. Kanbur. Oxford, Oxford University Press: 83–99.

Schelling, T. C. (1969). "Models of segregation." *American Economic Review* 59(2): 488–493.

Scheuer, C., G. A. Keoleian, et al. (2003). "Life cycle energy and environmental performance of a new university building: modeling challenges and design implications." *Energy and Buildings* 35(10): 1049–1064.

Schipper, L. (2004). "International Comparisons of Energy End Use: Benefits and Risks." *Encyclopedia of Energy*. C. J. Cleveland. Amsterdam, the Netherlands, Elsevier. 3: 529–555.

Schneider, A., M. A. Friedl and D. Potere (2009). "A new map of global urban extent from MODIS satellite data." *Environmental Research Letters* 4(4):044003 (11pp).

Scholz, Y. (2010). "Möglichkeiten und Grenzen der Integration Vershiedener Regnerativer Energiequellen zu einer 100% Regenerativen Stromversorgung der Bundesrepublik Deutschland bis zum Jahr 2050." *Endbericht für den Sachverständigenrat für Umweltfragen*. Cologne, Germany, German Aerospace Centre.

Scholz, Y. (2011). Data based on "Renewable Energy Based Electricity Supply at Low Costs – Setup and Application of the REMix model for Europe." Stuttgart, Germany, German Aerospace Centre.

Schulz, N. B. (2005). "Contributions to Material and Energy Flow Accounting to Urban Ecosystems Analysis." UNU-IAS *Working Paper No. 136*. Yokohama, Japan, United Nations University, Institute of Advanced Studies.

Schulz, N. B. (2006). "Socio-economic Development and Society's Metabolism in Singapore." *UNU-IAS Working Paper No. 148*. Yokohama, Japan, United Nations University, Institute of Advanced Studies.

Schulz, N. B. (2007). "The direct material inputs into Singapore's development." *Journal of Industrial Ecology* 11(2): 117–131.

Schulz, N. B. (2010a). "Delving into the carbon footprints of Singapore – Comparing direct and indirect greenhouse gas emissions of a small and open economic system." *Energy Policy* 38(9): 4848–4855.

Schulz, N. B. (2010b). "Urban Energy Consumption Database and Estimations of Urban Energy Intensities." Laxenburg, Austria, International Institute for Applied Systems Analysis (IIASA).

Schwela, D., G. Haq, et al. (2006). *Urban Air Pollution in Asian Cities: Status, challenges and management*. London, UK, Earthscan.

Seto, K. C. and J. M. Shepherd (2009). "Global urban land-use trends and climate impacts." *Current Opinion in Environmental Sustainability* 1(1): 89–95.

Shrestha, R. M., S. Kumar, et al. (2008). "Modern energy use by the urban poor in Thailand: A study of slum households in two cities." *Energy for Sustainable Development* 12(4): 5–13.

Smil, V. (1991). *General Energetics. Energy in the biosphere and civilizations*. New York, NY, USA, John Wiley & Sons.

Smith, K. R. (1993). "Fuel combustion, air pollution exposure, and health: The situation in developing countries." *Annual Review of Energy and the Environment* 18(1): 529–566.

Sornette, D. and R. Cont (1997). "Convergent multiplicative processes repelled from zero: Power laws and truncated power laws." *Journal de Physique I* 7: 431–444.

Souch, C. and S. Grimmond (2006). "Applied climatology: Urban climate." *Progress in Physical Geography* 30(2): 270–279.

SPARC (1990). "The Society for the Promotion of Area Resource Centres (SPARC): Developing new NGO lines." *Environment and Urbanization* 2(1): 91–104.

Stead, D. and J. Williams (2000). "Land use, Transport and People: Identifying the Connections." *Achieving Sustainable Urban Form*. K. Williams, E. Burton and M. Jenks. London, UK, Spon Press, Taylor and Francis.

Steemers, K. (2003). "Energy and the city: Density, buildings and transport." *Energy and Buildings* 35(1): 3–14.

Steingrube, W. and R. Boerdlein (2009). "Greifswald – die Fahrradstadt, Ergebnisse der Befragung zur Verkehrsmittel der Greifswalder Bevölkerung 2009" (Greifswald – The Bicycle City, Results of a survey for the modal split of the Greifswald population). Greifswald, Germany, Institut für Geographie und Geologie der Universität Greifswald.

Stephens, C., P. Rajesh, et al. (1996). "This is My Beautiful Home: Risk Perceptions towards Flooding and Environment in Low Income Urban

Communities: A Case Study in Indore, India." London, UK, London School of Hygiene and Tropical Medicine: 51.

Stern, D. I. (2004). "The rise and fall of the environmental Kuznets curve." *World Development* 32(8): 1419–1439.

Supernak, J. (2005). *HOT Lanes on Interstate 15 in San Diego: Technology, Impacts and Equity Issues*. PIARC Seminar on Road Pricing with Emphasis on Financing, Regulation and Equity, Cancun, Mexico.

Taha, H. (1997). "Urban climate and heat islands: Albedo, evapotranspiration and anthropogenic heat." *Energy and Buildings* (25): 99–103.

Tarr, J. (2001). Urban history and environmental history in the United States: Complementary and overlapping fields. *Environmental Problems in European Cities of the 19th and 20th Century*. C. Bernhardt. New York, Munich, Berlin, Waxmann, Muenster: 25–39.

Tarr, J. (2005). "The city and technology." *A Companion to American Technology*. C. Pursell. New York, Blackwell Publishing: 97–112.

Tatiétsé, T. T., P. Villeneuve, et al. (2002). "Contribution to the analysis of urban residential electrical energy demand in developing countries." *Energy* 27(6): 591–606.

Taylor, B. D. (2004). "The politics of congestion mitigation." *Transport Policy* 11(3): 299–302.

Tfl (2007). "London freight plan: Sustainable freight distribution: A plan for London." from www.tfl.gov.uk/microsites/freight/documents/London-Freight-Plan.pdf.

Tokyo Metropolitan Government (2006). "Environmental White paper 2006." Tokyo, Japan, Tokyo Metropolitan Government.

Treberspurg, M. (2005). "Nachhaltige und zukunftssichere Architektur durch ressourcenorientiertes Bauen." *Österreichische Wasser- und Abfallwirtschaft* 57(7): 111–117.

Uchida, H., and A. Nelson (2008). "Agglomeration Index: Towards a New Measure of Urban Concentration." *World Development Report Background Paper*, Washington DC, USA, World Bank.

UKDECC (2010). "Annual Report on Fuel Poverty Statistics 2009." London, UK, United Kingdom Department of Energy and Climate Change (UKDECC).

Umweltzone Berlin (2009). "Aktuelles Umweltzone." Retrieved 26 January, 2010, from www.berlin.de/Umweltzone.

UN (2008). "Report of the Secretary-General: World population monitoring, focusing on population distribution, urbanization, internal migration and development." *United Nations (UN) Commission on Population and Development 41st Session*. New York, NY, USA, United Nations.

UN (2009). "World Urbanization Prospects." New York.

UN DESA (1971). "Some simple methods for urban and rural population forecasts." New York, United Nations Department of Economic and Social Affairs (UN DESA).

UN DESA (1973). *The determinants and consequences of population trends: new summary of findings on interaction of demographic, economic and social factors*. New York, United Nations Department of Economic and Social Affairs (UN DESA).

UN DESA (1974). "Manual VIII: Methods for Projections of Urban and Rural Population." New York, United Nations Department of Economic and Social Affairs (UN DESA).

UN DESA (1980). "Patterns of Urban and Rural Population Growth." *Population Studies*. New York, Population Division, United Nations Department of Economic and Social Affairs.

UN DESA (2010). "World Urbanization Prospects: The 2009 Revision." New York, Population Division, United Nations Department of Economic and Social Affairs (UN DESA).

UN HABITAT (2003). *The Challenge of Slums. Global Report on Human Settlements 2003*. London, UK, United Nations Human Settlements Programme (UN-HABITAT), Earthscan.

UN HABITAT (2008). *State of the World's Cities 2008/2009. Harmonious cities*. London, UK, United Nations Human Settlements Programme (UN-HABITAT), Earthscan.

UNDP and WHO (2009). "The Energy Access Situation in Developing Countries: A Review Focusing on the Least Developed Countries and Sub-Saharan Africa." New York, NY, USA.

UNEP (2009). "CNG conversion: Learning from New Delhi." Retrieved 2 November, 2009, from ekh.unep.org/?q=node/1737.

UNEP (2011). "Decoupling natural resource use and environmental impacts from economic growth." United Nations Environmental Programme.

UNIDO (2009). "Breaking In and Moving Up: New Industrial Challenges for the Bottom Billion and the Middle-Income Countries." *Industrial Development Report 2009*. Vienna, Austria, United Nations Industrial Development Organization (UNIDO): 146.

UNITE (2003). "Unification of accounts and marginal costs for Transport Efficiency." Leeds, UK, European Commission under the Transport RTD Programme of the 5th Framework Programme, the University of Leeds.

Urban Audit (2009). Retrieved 30 December, 2009, from www.urban audit.org.

Urban Resource Centre (2001). "Urban poverty and transport: A case study from Karachi." *Environment and Urbanization* 13(1): 223–233.

US DOE (2005). "Residential Energy Consumption Survey." Retrieved 19 September, 2010, from eia.doe.gov/emeu/recs/recs2005/hc2005_tables/detailed_tables2005.html.

USAID (2004). "Innovative Approaches to Slum Electrification." Washington, DC, USA, Bureau for Economic Growth, Agriculture and Trade; United States Agency for International Development (USAID).

USEPA (2009). "National trends in sulfur dioxide levels." Retrieved 11 May, 2009, from www.epa.gov/airtrends/sulfur.html.

van der Plas, R. J. and M. A. Abdel-Hamid (2005). "Can the woodfuel supply in sub-Saharan Africa be sustainable? The case of N'Djaména, Chad." *Energy Policy* 33(3): 297–306.

VandeWeghe, J. R. and C. Kennedy (2007). "A spatial analysis of residential greenhouse gas emissions in the Toronto census metropolitan area." *Journal of Industrial Ecology* 11(2): 133–144.

Viswanathan, B. and K. S. Kavi Kumar (2005). "Cooking fuel use patterns in India: 1983–2000." *Energy Policy* 33(8): 1021–1036.

Vivier, J. (2006). "Mobility in Cities Database, Better Mobility for People Worldwide, Analysis and Recommendations." Brussels, Belgium, International Association of Public Transport (UITP).

Vringer, K. and K. Blok (1995). "The direct and indirect energy requirements of households in the Netherlands." *Energy Policy* 23(10): 893–910.

Wapedia (2009). Retrieved December 30, 2009, from wapedia.mobi/en/Modal_share.

Watkiss, P., C. Brand, et al. (2000). "Urban Energy-Use: Guidance on Reducing Environmental Impacts." *DFID KaR Project R7369*. Abingdon, UK, United Kingdom Department for International Development (DfID).

WCED (1987). *Our Common Future, Report of the World Commission on Environment and Development (WCED)*. New York and Oxford, UK, World Commission on Environment and Development (WCED), Oxford University Press.

Weber, C. and N. Shah (2011). "Optimisation based design of a district's energy system for an eco-town in the United Kingdom." *Energy* 36(2): 1292–1308.

Weber, C. L. and H. S. Matthews (2008). "Quantifying the global and distributional aspects of American household carbon footprint." *Ecological Economics* 66(2–3): 379–391.

WEC (2006). *Alleviating Urban Energy Poverty in Latin America*. London, UK, World Energy Council (WEC).

Weisz, H. and F. Duchin (2006). "Physical and monetary input–output analysis: What makes the difference?" *Ecological Economics* 57(3): 534–541.

Weisz, H., F. Krausmann, et al. (2006). "The physical economy of the European Union: Cross-country comparison and determinants of material consumption." *Ecological Economics* 58(4): 676–698.

Weng, Q., D. Lu, et al. (2004). "Estimation of land surface temperature–vegetation abundance relationship for urban heat island studies." *Remote Sensing of Environment* 89(4): 467–483.

West, G. B., J. H. Brown, et al. (1999). "The fourth dimension of life: Fractal geometry and allometric scaling of organisms." *Science* 284(5420): 1677–1679.

WHO (2002). *World Health Report 2002: Reducing risk, promoting healthy life*. Geneva, Switzerland, World Health Organization (WHO).

WHO (2006). *Fuel for Life: Household energy and health*. Geneva, Switzerland, World Health Organization (WHO).

Wiechmann, T. (2009). "Conversion strategies under uncertainty in post-socialist shrinking cities: the example of Dresden in Eastern Germany." *The Future of Shrinking Cities: Problems, Patterns and Strategies of Urban Transformation in a Global Context*. K. Pallagst, J. Aber and I. Audirac. Berkeley, CA, USA, Center for Global Metropolitan Studies, Institute of Urban and Regional Development: 5–15.

Wiedenhofer, D., M. Lenzen, and J. K. Steinberger (2011). "Spatial and socio-economic drivers of direct and indirect household energy consumption in Australia." *Urban Consumption*. P.W. Newton (ed), CISRO Publishing, Collingwood, Australia, pp. 251–266.

Wiedmann, T., R. Wood, et al. (2007). "Development of an Embedded Carbon Emissions Indicator – Producing a Time Series of Input–Output Tables and Embedded Carbon Dioxide Emissions for the UK by Using a MRIO Data Optimisation System." *Report to the UK Department for Environment, Food and Rural Affairs by Stockholm Environment Institute at the University of York and Centre for Integrated Sustainability Analysis at the University of Sydney*. London, UK, DEFRA.

Wier, M., M. Lenzen, et al. (2001). "Effects of household consumption patterns on CO_2 requirements." *Economic Systems Research* 13: 259–274.

Wilson, A. G. (2000). *Complex Spatial Systems: The modelling foundations of urban and regional analysis*. Upper Saddle River, NJ, USA, Prentice Hall.

Winrock International (2005). "Enabling Urban Poor Livelihoods Policy Making: Understanding the Role of Energy Services – Country Report for the Brazil Project." *DFID KaR Project R8348*. Salvador, Brazil, Winrock International and the United Kingdom Department for International Development (DfID).

Wolman, A. (1965). "The metabolism of cities." *Scientific American* 213(3): 178–193.

World Bank (1992). *World Development Report*. Washington, DC, USA, World Bank.

World Bank (2009). *The World Development Report 2009: Reshaping Economic Geography*. Washington, DC, USA, World Bank.

Wu, X., J. Lampietti, et al. (2004). "Coping with the cold: Space heating and the urban poor in developing countries." *Energy Economics* 26(3): 345–357.

Yamagata, Y. (2010). Personal communication of detailed data set of Kinoshita et al., 2008. Tsukuba, Japan, National Institute for Environmental Studies (NIES).

Yamaguchi, Y., Y. Shimoda, et al. (2007). "Transition to a sustainable urban energy system from a long-term perspective: Case study in a Japanese business district." *Energy and Buildings* 39(1): 1–12.

Yapi-Diahou, A. (1995). "The informal housing sector in the metropolis of Abdjan, Ivory Coast." *Environment and Urbanization* 7(2):11–30.

Zahavi, Y. (1974). "Traveltime Budgets and Mobility in Urban Areas." Washington DC, USA, United States Department of Transportation: 49.

Zhou, H. and D. Sperling (2001). "Transportation in Developing Countries. Greenhouse Gas Scenarios for Shanghai, China." Arlington, VA, USA, Pew Center: 43.

Zhou, P. (2000). "Energy Use in Urban Transport in Africa." *Energy Environment Linkages in African Cities*. Nairobi, Kenya, United Nations Centre for Settlements (Habitat) and the United Nations Environment Programme (UNEP): 100–111.

Zipf, G. (1949). *Human Behavior and the Principle of Least-Effort: An introduction to human ecology*. Cambridge, MA, USA, Addison-Wesley.

Index

Please note that page numbers relating to Notes will have the letter 'n' following the page number. References to Figures or Tables will be in *italics*.